Numerical Integration of Space Fractional Partial Differential Equations

Vol 2 - Applications from Classical Integer PDEs

Synthesis Lectures on Mathematics and Statistics

Editor
Steven G. Krantz, *Washington University, St. Louis*

Essentials of Applied Mathematics for Engineers and Scientists, Second Edition
Robert G. Watts
2012

Chaotic Maps: Dynamics, Fractals, and Rapid Fluctuations
Goong Chen and Yu Huang
2011

Matrices in Engineering Problems
Marvin J. Tobias
2011

The Integral: A Crux for Analysis
Steven G. Krantz
2011

Statistics is Easy! Second Edition
Dennis Shasha and Manda Wilson
2010

Lectures on Financial Mathematics: Discrete Asset Pricing
Greg Anderson and Alec N. Kercheval
2010

Jordan Canonical Form: Theory and Practice
Steven H. Weintraub
2009

The Geometry of Walker Manifolds
Miguel Brozos-Vázquez, Eduardo García-Río, Peter Gilkey, Stana Nikčević, and Ramón Vázquez-Lorenzo
2009

An Introduction to Multivariable Mathematics
Leon Simon
2008

Jordan Canonical Form: Application to Differential Equations
Steven H. Weintraub
2008

Statistics is Easy!
Dennis Shasha and Manda Wilson
2008

A Gyrovector Space Approach to Hyperbolic Geometry
Abraham Albert Ungar
2008

Numerical Integration of Space Fractional Partial Differential Equations:
Vol 2 - Applications from Classical Integer PDEs
Younes Salehi and William E. Schiesser

ISBN: 978-3-031-01284-6   paperback
ISBN: 978-3-031-02412-2   ebook

DOI 10.1007/978-3-031-02412-2

A Publication in the Springer series
SYNTHESIS LECTURES ON MATHEMATICS AND STATISTICS

Lecture #20
Series Editor: Steven G. Krantz, *Washington University, St. Louis*
Series ISSN
Print 1938-1743   Electronic 1938-1751

Numerical Integration of Space Fractional Partial Differential Equations

Vol 2 - Applications from Classical Integer PDEs

Younes Salehi
Razi University

William E. Schiesser
Lehigh University

SYNTHESIS LECTURES ON MATHEMATICS AND STATISTICS #20

ABSTRACT

Partial differential equations (PDEs) are one of the most used widely forms of mathematics in science and engineering. PDEs can have partial derivatives with respect to (1) an initial value variable, typically time, and (2) boundary value variables, typically spatial variables. Therefore, two fractional PDEs can be considered, (1) fractional in time (TFPDEs), and (2) fractional in space (SFPDEs). The two volumes are directed to the development and use of SFPDEs, with the discussion divided as:

- Vol 1: Introduction to Algorithms and Computer Coding in R

- Vol 2: Applications from Classical Integer PDEs.

Various definitions of space fractional derivatives have been proposed. We focus on the *Caputo* derivative, with occasional reference to the *Riemann-Liouville* derivative.

In the second volume, the emphasis is on applications of SFPDEs developed mainly through the extension of classical integer PDEs to SFPDEs. The example applications are:

- Fractional diffusion equation with Dirichlet, Neumann and Robin boundary conditions

- Fisher-Kolmogorov SFPDE

- Burgers SFPDE

- Fokker-Planck SFPDE

- Burgers-Huxley SFPDE

- Fitzhugh-Nagumo SFPDE

These SFPDEs were selected because they are integer first order in time and integer second order in space. The variation in the spatial derivative from order two (parabolic) to order one (first order hyperbolic) demonstrates the effect of the spatial fractional order α with $1 \le \alpha \le 2$. All of the example SFPDEs are one dimensional in Cartesian coordinates. Extensions to higher dimensions and other coordinate systems, in principle, follow from the examples in this second volume.

The examples start with a statement of the integer PDEs that are then extended to SFPDEs. The format of each chapter is the same as in the first volume.

The R routines can be downloaded and executed on a modest computer (R is readily available from the Internet).

KEYWORDS

partial differential equations, value variables; space fractional partial differential equations, fractional calculus

To

Mahnaz and Dolores

for their encouragement and patience.

Contents

Preface

Partial differential equations (PDEs) are one of the most used widely forms of mathematics in science and engineering. PDEs can have partial derivatives with respect to (1) an initial value variable, typically time, and (2) boundary value variables, typically spatial variables. Therefore, two fractional PDEs (FPDEs) can be considered, (1) fractional in time (TFPDEs), and (2) fractional in space (SFPDEs). The books[1] are directed to the development and use of SFPDEs.

FPDEs have features and solutions that go beyond the established integer PDEs (IPDEs), for example, the classical field equations including the Euler, Navier-Stokes, Maxwell and Einstein equations. FPDEs therefore offer the possibility of solutions that have features that better approximate physical/chemical/biological phenomena than IPDEs.

Fractional calculus dates back to the beginning of calculus (e.g., to Leibniz, Riemann and Liouville), but recently there has been extensive reporting of applications, typically as expressed by TFPDE/SFPDEs. In particular, SFPDEs are receiving broad attention in the research literature, especially when applied to the computer-based modeling of heterogeneous media. For example, SFPDEs are being applied to living tissue (with potential applications in biomedical engineering, biology and medicine).

Various definitions of space fractional derivatives have been proposed. Therefore, as a first step in the use of SFPDEs, a definition of the derivative must be selected. In the books, we focus on the *Caputo* derivative, with occasional reference to the *Riemann-Liouville* derivative.

The Caputo derivative has at least two important advantages:

1. For the special case of an integer derivative, the usual properties of integer calculus follow. For example, the Caputo derivative of a constant is zero.

2. The definition of a Caputo derivative is based the integral of an integer derivative. Therefore, the established algorithms for approximating integer derivatives can be used. For the numerical methods that follow, the integer derivatives are approximated with splines.

The Caputo derivative is defined as a convolution integral. Thus, rather than being *local* (with a value at a particular point in space), the Caputo derivative is *non-local*, (it is based on an integration in space), which is one of the reasons that it has properties not shared by integer derivative).

---

[1]The two volume set has the titles:
Numerical Integration of Space Fractional Partial Differential Equations
Vol 1: Introduction to Algorithms and Computer Coding in R
Vol 2: Applications from Classical Integer PDEs.

A parameter of the Caputo derivative that is of primary interest is the order of the derivative, which is fractional, with integer order as a special case. The various example applications that follow generally permit the variation of the fractional order in computer-based analysis.

The papers cited as a source of the SFPDE models generally consist of a statement of the equations followed by reported numerical solutions. Generally, little or no information is given about how the solutions were computed (the algorithms) and in all cases, the computer code that was used to calculate the solutions is not provided.

In other words, what is missing is: (1) a detailed discussion of the numerical methods used to produce the reported solutions and (2) the computer routines used to calculate the reported solutions. For the reader to complete these two steps to verify the reported solutions with reasonable effort is essentially impossible.

A principal objective of the books is therefore to provide the reader with a set of documented R routines that are discussed in detail, and can be downloaded and executed without having to first master the details of the relevant numerical analysis and then code a set of routines.

The example applications are intended as introductory and open ended. They are based mainly on classical (legacy) IPDEs. The focus in each chapter is on:

1. A statement of the SFPDE system, including initial conditions (ICs), boundary conditions (BCs) and parameters.

2. The algorithms for the calculation of numerical solutions, with particular emphasis on splines.

3. A set of R routines for the calculation of numerical solutions, including a detailed explanation of each section of the code.

4. Discussion of the numerical solution.

5. Summary and conclusions about extensions of the computer-based analysis.

In summary, the presentation is not as formal mathematics, e.g., theorems and proofs. Rather, the presentation is by examples of SFPDE applications, including the details for computing numerical solutions, particularly with documented source code. The authors would welcome comments, especially pertaining to this format and experiences with the use of the R routines. Comments and questions can be directed to wes1@lehigh.edu.

Younes Salehi and William E. Schiesser
November 2017

CHAPTER 6

Simultaneous SFPDEs

6.1 INTRODUCTION

The preceding examples in Volume 1 demonstrate the MOL solution of a fractional PDE (e.g., eqs. (1.3a), (1.4a)). In this chapter we consider an example of simultaneous SFPDEs which model fractional diffusion and transfer that connects the two SFPDEs.

6.2 SIMULTANEOUS SFPDES

The simultaneous SFPDEs are

$$\frac{\partial u_1}{\partial t} = \frac{\partial^\alpha u_1}{\partial x^\alpha} + k_{12}(u_2 - u_1) \tag{6.1a}$$

$$\frac{\partial u_2}{\partial t} = \frac{\partial^\beta u_2}{\partial x^\beta} - k_{12}(u_2 - u_1) \tag{6.1b}$$

with $1 \leq \alpha, \beta \leq 2$. k_{12} is a coefficient that can represent, for example, mass transfer connecting two regions with concentrations $u_1(x,t), u_2(x,t)$. As a variant, $k_2 u_2 - k_1 u_1$ in eq. (6.1a) could represent a net first order reaction rate. A similar term would then also appear in eq. (6.1b). Also, the rates of reaction could be nonlinear, e.g., $k_2 u_2^2 - k_1 u_1^2$ or $k_{12} u_1^a u_2^b - k_{21} u_1^c u_2^d$. These various forms of the transfer/reaction rates can be readily accommodated within the numerical implementation to follow.

Eqs. (6.1) are first order in t so that each requires one initial condition (IC).

$$u_1(x, t = 0) = f_1(t); \quad u_2(x, t = 0) = f_2(t) \tag{6.2a,b}$$

As discussed with previous examples, eqs. (6.1) each require two boundary conditions (BCs). Homogeneous Dirichlet BCs are considered first.

$$u_1(x = x_l, t) = u_1(x = x_u, t) = 0 \tag{6.3a,b}$$

$$u_2(x = x_l, t) = u_2(x = x_u, t) = 0 \tag{6.3c,d}$$

Then, homogeneous Neumann BCs are considered.

$$\frac{\partial u_1(x = x_l, t)}{\partial x} = \frac{\partial u_1(x = x_u, t)}{\partial x} = 0 \tag{6.4a,b}$$

$$\frac{\partial u_2(x = x_l, t)}{\partial x} = \frac{\partial u_2(x = x_u, t)}{\partial x} = 0 \qquad (6.4c,d)$$

Finally, Robin BCs are considered.

$$c_2 \frac{\partial u_1(x = x_l, t)}{\partial x} + c_1 u_1(x = x_l, t) = 0 \qquad (6.5a)$$

$$c_4 \frac{\partial u_1(x = x_u, t)}{\partial x} + c_3 u_1(x = x_u, t) = 0 \qquad (6.5b)$$

$$c_2 \frac{\partial u_2(x = x_l, t)}{\partial x} + c_1 u_2(x = x_l, t) = 0 \qquad (6.5c)$$

$$c_4 \frac{\partial u_2(x = x_u, t)}{\partial x} + c_3 u_2(x = x_u, t) = 0 \qquad (6.5d)$$

Eqs. (6.1), (6.2), (6.3) constitute the initial SFPDE system now analyzed by the method of lines (MOL).

6.2.1   MAIN PROGRAM

A main program for eqs. (6.1), (6.2), (6.3) follows.

Listing 6.1: Main program for eqs. (6.1), (6.2), (6.3)

```
#
# Simultaneous SFPDEs
#
#   u1t=d^alpha u1/dx^alpha+k12*(u2-u1)
#
#   u2t=d^beta  u2/dx^beta -k12*(u2-u1)
#
#   xl < x < xu, 0 < t < tf, xl=0, xu=1
#
#   u1(x,t=0)=e^(-100*(x-0.5)^2)
#
#   u2(x,t=0)=0
#
#   u1(x=xl,t)=u1(x=xu,t)=0
#
#   u2(x=xl,t)=u2(x=xu,t)=0
#
# Delete previous workspaces
  rm(list=ls(all=TRUE))
#
```

```
# Access functions for numerical solution
  library("deSolve");
  setwd("f:/fractional/sfpde/chap6");
  source("pde1a.R");
#
# Parameters
  ncase=1;
  if(ncase==1){
    alpha=1.5;beta=1.5;k12=0;
    c_1=1;c_2=0;c_3=1;c_4=0;}
#
# Initial condition function (IC)
  f1=function(x) exp(-100*(x-0.5)^2);
  f2=function(x) 0;
#
# Boundary condition functions (BCs)
  g_0=function(t) 0;
  g_L=function(t) 0;
#
# Spatial grid
  xl=0;xu=1;nx=51;dx=(xu-xl)/(nx-1);
  xj=seq(from=xl,to=xu,by=dx);
  cd=dx^(-alpha)/gamma(4-alpha);
#
# Independent variable for ODE integration
  t0=0;tf=0.1;nt=6;dt=(tf-t0)/(nt-1);
  tout=seq(from=t0,to=tf,by=dt);
#
# a1_jk coefficients
  A1=matrix(0,nrow=nx-2,ncol=nx-1);
  for(j in 1:(nx-2)){
    for(k in 0:j){
    if (k==0){
      A1[j,k+1]=(j-1)^(3-alpha)-j^(2-alpha)*(j-3+alpha);
    } else if (1 <= k && k<=j-1){
      A1[j,k+1]=(j-k+1)^(3-alpha)-2*(j-k)^(3-alpha)+(j-k-1)^(3-
        alpha);
    } else
      A1[j,k+1]=1;
```

```
      }
    }
#
# a2_jk coefficients
  A2=matrix(0,nrow=nx-2,ncol=nx-1);
  for(j in 1:(nx-2)){
    for(k in 0:j){
    if (k==0){
      A2[j,k+1]=(j-1)^(3-beta)-j^(2-beta)*(j-3+beta);
    } else if (1 <= k && k<=j-1){
      A2[j,k+1]=(j-k+1)^(3-beta)-2*(j-k)^(3-beta)+(j-k-1)^(3-beta
        );
    } else
      A2[j,k+1]=1;
    }
  }
#
# Initial condition
  nx=nx-2;
  u0=rep(0,2*nx);
  for(j in 1:nx){
    u0[j]    =f1(xj[j+1]);
    u0[j+nx]=f2(xj[j+1]);
}
  ncall=0;
#
# ODE integration
  out=lsode(y=u0,times=tout,func=pde1a,
      rtol=1e-6,atol=1e-6,maxord=5);
  nrow(out)
  ncol(out)
#
# Allocate arrays for u1(x,t), u2(x,t)
  nx=nx+2;
  u1=matrix(0,nt,nx);
  u2=matrix(0,nt,nx);
#
# u1(x,t), u2(x,t), x ne xl,xu
  for(i in 1:nt){
```

```
    for(j in 2:(nx-1)){
      u1[i,j]=out[i,j];
      u2[i,j]=out[i,j+(nx-2)];
    }
  }
#
# Reset boundary values
  for(i in 1:nt){
   u1[i,1]=g_0(tout[i]);
  u1[i,nx]=g_L(tout[i]);
   u2[i,1]=g_0(tout[i]);
  u2[i,nx]=g_L(tout[i]);
  }
#
# Tabular numerical solutions
  cat(sprintf("\n\n   alpha = %4.2f  beta = %4.2f\n",
              alpha,beta));
  cat(sprintf("\n      t     x    u1(x,t)   u2(x,t)"));
  for(i in 1:nt){
  iv=seq(from=1,to=nx,by=5);
  for(j in iv){
    cat(sprintf("\n %6.2f%6.2f%10.5f%10.5f",
      tout[i],xj[j],u1[i,j],u2[i,j]));
  }
  cat(sprintf("\n"));
  }
#
# Plot numerical u1 solution
  matplot(xj,t(u1),type="l",lwd=2,col="black",lty=1,
    xlab="x",ylab="u1(x,t)",main="");
#
# Plot numerical u2 solution
  matplot(xj,t(u2),type="l",lwd=2,col="black",lty=1,
    xlab="x",ylab="u2(x,t)",main="");
#
# Calls to ODE routine
  cat(sprintf("\n\n   ncall = %3d\n",ncall));
```

We can note the following details about Listing 6.1.

- Brief documentation comments are followed by deletion of previous files.

```
#
# Simultaneous SFPDEs
#
#    u1t=d^alpha u1/dx^alpha+k12*(u2-u1)
#
#    u2t=d^beta  u2/dx^beta -k12*(u2-u1)
#
#    xl < x < xu, 0 < t < tf, xl=0, xu=1
#
#    u1(x,t=0)=e^(-100*(x-0.5)^2)
#
#    u2(x,t=0)=0
#
#    u1(x=xl,t)=u1(x=xu,t)=0
#
#    u2(x=xl,t)=u2(x=xu,t)=0
#
# Delete previous workspaces
  rm(list=ls(all=TRUE))
```

- The ODE integrator library deSolve is accessed. Note that the setwd (set working directory) uses / rather than the usual \.

```
#
# Access functions for numerical solution
  library("deSolve");
  setwd("f:/fractional/sfpde/chap6");
  source("pde1a.R");
```

- The parameters for ncase=1 are defined,

```
#
# Parameters
  ncase=1;
  if(ncase==1){
    alpha=1.5;beta=1.5;k12=0;
    c_1=1;c_2=0;c_3=1;c_4=0;}
```

In particular, the orders of the fractional derivatives and the coupling coefficient in eqs. (6.1), α, β, k_{12}, are defined. Then the coefficients are defined for the homogeneous

Dirichlet BCs of eqs. (6.3) (see eqs. (6.5)). These coefficients are used in the ODE/MOL routine pde1a discussed subsequently.

Additional cases are subsequently formulated as they are first discussed.

- The IC functions of eqs. (6.2) are defined.

```
#
# Initial condition function (IC)
  f1=function(x) exp(-100*(x-0.5)^2);
  f2=function(x) 0;
```

- BC functions, used in pde1a, are defined.

```
#
# Boundary condition functions (BCs)
  g_0=function(t) 0;
  g_L=function(t) 0;
```

- A spatial grid of 51 points is defined for the interval $x = x_l = 0 \le x \le x = x_u = 1$, so that grid values are $x = 0, 0.02, ..., 1$.

```
#
# Spatial grid
  xl=0;xu=1;nx=51;dx=(xu-xl)/(nx-1);
  xj=seq(from=xl,to=xu,by=dx);
  cd=dx^(-alpha)/gamma(4-alpha);
```

cd is used in the ODE/MOL routine pde1a.

- The interval in t, $t = t_0 = 0 \le t \le t = t_f = 0.1$, is defined with 6 points, so the output values are $t = 0, 0.02, ..., 0.1$.

```
#
# Independent variable for ODE integration
  t0=0;tf=0.1;nt=6;dt=(tf-t0)/(nt-1);
  tout=seq(from=t0,to=tf,by=dt);
```

- The coefficient matrix of eq. (1.2g) is defined for eq. (6.1a).

```
#
# a1_jk coefficients
  A1=matrix(0,nrow=nx-2,ncol=nx-1);
```

```
    for(j in 1:(nx-2)){
      for(k in 0:j){
      if (k==0){
        A1[j,k+1]=(j-1)^(3-alpha)-j^(2-alpha)*(j-3+alpha);
      } else if (1 <= k && k<=j-1){
        A1[j,k+1]=(j-k+1)^(3-alpha)-2*(j-k)^(3-alpha)+(j-k-1)^(3-alpha);
      } else
        A1[j,k+1]=1;
      }
    }
```

Note the use of α alpha.

- The coefficient matrix of eq. (1.2g) is defined for eq. (6.1b).

```
#
# a2_jk coefficients
  A2=matrix(0,nrow=nx-2,ncol=nx-1);
  for(j in 1:(nx-2)){
    for(k in 0:j){
    if (k==0){
      A2[j,k+1]=(j-1)^(3-beta)-j^(2-beta)*(j-3+beta);
    } else if (1 <= k && k<=j-1){
      A2[j,k+1]=(j-k+1)^(3-beta)-2*(j-k)^(3-beta)+(j-k-1)^(3-beta);
    } else
      A2[j,k+1]=1;
    }
  }
```

Note the use of β beta.

- The ICs, eqs. (6.2), for ncase=1 are taken as the same Gaussian function $f_1(x,t)$ for $u_1(x,t)$ and $u_2(x,t)$. For this repeated IC, the solution should also be repeated, which can be used as a check of the coding, i.e., a difference in the solutions for $u_1(x,t)$ and $u_2(x,t)$ would indicate a programming error.

```
#
# Initial condition
  nx=nx-2;
  u0=rep(0,2*nx);
  for(j in 1:nx){
```

```
    u0[j]   =f1(xj[j+1]);
    u0[j+nx]=f1(xj[j+1]);
}
  ncall=0;
```

The counter for the calls to the ODE/MOL routine pde1a is also initialized.

- The $2(51 - 2) = 2(49) = 98$ ODEs programmed in pde1a, corresponding to the 49 interior points in x, are integrated by lsode.

```
#
# ODE integration
  out=lsode(y=u0,times=tout,func=pde1a,
      rtol=1e-6,atol=1e-6,maxord=5);
  nrow(out)
  ncol(out)
```

The dimensions of the solution array, out, are $6 \times 98 + 1 = 99$ as confirmed in the numerical output that follows. The offset of $+1$ indicates that the value of t is included as the first element of each solution vector.

- Arrays u1,u2 are allocated for the solutions $u_1(x, t), u_2(x, t)$ (nt=6, nx=51).

```
#
# Allocate arrays for u1(x,t), u2(x,t)
  nx=nx+2;
  u1=matrix(0,nt,nx);
  u2=matrix(0,nt,nx);
```

- The 98 ODE solutions are placed in u1,u2.

```
#
# u1(x,t), u2(x,t), x ne xl,xu
  for(i in 1:nt){
    for(j in 2:(nx-1)){
      u1[i,j]=out[i,j];
      u2[i,j]=out[i,j+(nx-2)];
    }
  }
```

The for in j starts at j=2 since j=1 corresponds to the value of t (at each i).

- The boundary values $u_1(x = x_l, t) = u_1(x = x_u, t) = 0$, $u_2(x = x_l, t) = u_2(x = x_u, t) = 0$ are added to u1,u2.

```
#
# Reset boundary values
  for(i in 1:nt){
   u1[i,1]=g_0(tout[i]);
  u1[i,nx]=g_L(tout[i]);
   u2[i,1]=g_0(tout[i]);
  u2[i,nx]=g_L(tout[i]);
  }
```

- Numerical values of the solutions $u_1(x,t), u_2(x,t)$ are displayed. Every fifth value of x appears from by=5.

```
#
# Tabular numerical solutions
  cat(sprintf("\n\n   alpha = %4.2f  beta = %4.2f\n",
             alpha,beta));
  cat(sprintf("\n     t     x    u1(x,t)   u2(x,t)"));
  for(i in 1:nt){
  iv=seq(from=1,to=nx,by=5);
  for(j in iv){
    cat(sprintf("\n %6.2f%6.2f%10.5f%10.5f",
      tout[i],xj[j],u1[i,j],u2[i,j]));
  }
  cat(sprintf("\n"));
  }
```

- $u_1(x,t)$ is plotted against x with t as a parameter. The number of rows of u1 transposed (t(u1)) equals the number of rows (elements) of xj.

```
#
# Plot numerical u1 solution
  matplot(xj,t(u1),type="l",lwd=2,col="black",lty=1,
    xlab="x",ylab="u1(x,t)",main="");
```

- Similarly, $u_2(x,t)$ is plotted against x with t as a parameter.

```
#
# Plot numerical u2 solution
```

```
  matplot(xj,t(u2),type="l",lwd=2,col="black",lty=1,
    xlab="x",ylab="u2(x,t)",main="");
```

- The number of calls to pde1a is displayed at the end of the solution.

```
#
# Calls to ODE routine
  cat(sprintf("\n\n   ncall = %3d\n",ncall));
```

The ODE/MOL routine pde1a is next.

6.2.2   ODE/MOL ROUTINE

Listing 6.2: ODE/MOL routine for eqs. (6.1), (6.2), (6.3)

```
pde1a=function(t,u,parms){
#
# Function pde1a computes the derivative
# vector of the ODEs approximating the
# PDE
#
# Place u in two vectors, u1, u2
  u1=rep(0,nx);
  u2=rep(0,nx);
  for(i in 1:nx){
    u1[i]=u[i];
    u2[i]=u[i+nx];
  }
#
# Allocate the vectors of the ODE
# derivatives
  u1t=rep(0,nx);
  u2t=rep(0,nx);
   ut=rep(0,2*nx);
#
# Approximation of u1xx
  u1xx=NULL;
#
# x=0
  u10=(2*dx*g_0(t)-c_2*(4*u1[1]-u1[2]))/
      (2*dx*c_1-3*c_2);
```

```
  u1xx_0=2*u10-5*u1[1]+4*u1[2]-u1[3];
  u1xx[1]=u1[2]-2*u1[1]+u10;
#
# x=1
  u1n=(2*dx*g_L(t)+c_4*(4*u1[nx]-u1[nx-1]))/
     (2*dx*c_3+3*c_4);
  u1xx[nx]=u1n-2*u1[nx]+u1[nx-1];
#
# Interior approximation of u1xx
  for(k in 2:(nx-1)){
    u1xx[k]=u1[k+1]-2*u1[k]+u1[k-1];
  }
#
# Approximation of u2xx
  u2xx=NULL;
#
# x=0
  u20=(2*dx*g_0(t)-c_2*(4*u2[1]-u2[2]))/
     (2*dx*c_1-3*c_2);
  u2xx_0=2*u20-5*u2[1]+4*u2[2]-u2[3];
  u2xx[1]=u2[2]-2*u2[1]+u20;
#
# x=1
  u2n=(2*dx*g_L(t)+c_4*(4*u2[nx]-u2[nx-1]))/
     (2*dx*c_3+3*c_4);
  u2xx[nx]=u2n-2*u2[nx]+u2[nx-1];
#
# Interior approximation of u2xx
  for(k in 2:(nx-1)){
    u2xx[k]=u2[k+1]-2*u2[k]+u2[k-1];
  }
#
# u1 PDE
#
# Step through ODEs
  for(j in 1:nx){
#
#   First term in series approximation of
#   fractional derivative
```

```
    u1t[j]=A1[j,1]*u1xx_0;
#
#    Subsequent terms in series approximation
#    of fractional derivative
    for(k in 1:j){
      u1t[j]=u1t[j]+A1[j,k+1]*u1xx[k];
#
#    Next k (next term in series)
    }
    u1t[j]=cd*u1t[j]+k12*(u2[j]-u1[j]);
#
# Next j (next ODE)
  }
#
# u2 PDE
#
# Step through ODEs
  for(j in 1:nx){
#
#    First term in series approximation of
#    fractional derivative
    u2t[j]=A2[j,1]*u2xx_0;
#
#    Subsequent terms in series approximation
#    of fractional derivative
    for(k in 1:j){
      u2t[j]=u2t[j]+A2[j,k+1]*u2xx[k];
#
#    Next k (next term in series)
    }
    u2t[j]=cd*u2t[j]-k12*(u2[j]-u1[j]);
#
# Next j (next ODE)
  }
#
# Place u1t, u2t in vector ut
  for(i in 1:nx){
    ut[i]    =u1t[i];
    ut[i+nx]=u2t[i];
```

```
   }
#
# Increment calls to pde1a
   ncall <<- ncall+1;
#
# Return derivative vector of ODEs
   return(list(c(ut)));
   }
```

We can note the following details about Listing 6.2.
- The function is defined.

```
   pde1a=function(t,u,parms){
#
# Function pde1a computes the derivative
# vector of the ODEs approximating the
# PDE
```

t is the current value of t in eqs. (6.1). u is the 98-vector of ODE/MOL dependent variables. parm is an argument to pass parameters to pde1a (unused, but required in the argument list). The arguments must be listed in the order stated to properly interface with lsode called in the main program of Listing 6.1. The derivative vector of the LHS of eqs. (6.1) is calculated next and returned to lsode.

- The vector u of ODE dependent variables is placed in two vectors, u1, u2, to facilitate the programming of eqs. (6.1).

```
#
# Place u in two vectors, u1, u2
   u1=rep(0,nx);
   u2=rep(0,nx);
   for(i in 1:nx){
     u1[i]=u[i];
     u2[i]=u[i+nx];
   }
```

- Vectors are allocated for the derivatives in t. Specifically, the 49-vectors of the derivatives $\frac{\partial u_1}{\partial t}, \frac{\partial u_2}{\partial t}$ are subsequently placed in arrays u1t, u2t, and the composite of these vectors is placed in ut to be returned to lsode.

```
#
# Allocate the vectors of the ODE
# derivatives
  u1t=rep(0,nx);
  u2t=rep(0,nx);
   ut=rep(0,2*nx);
```

- The fractional derivative $\dfrac{\partial^\alpha u_1}{\partial x^\alpha}$ in eq. (6.1a) is computed according to the algorithm of eqs. (1.2).

```
#
# Approximation of u1xx
  u1xx=NULL;
#
# x=0
  u10=(2*dx*g_0(t)-c_2*(4*u1[1]-u1[2]))/
      (2*dx*c_1-3*c_2);
  u1xx_0=2*u10-5*u1[1]+4*u1[2]-u1[3];
  u1xx[1]=u1[2]-2*u1[1]+u10;
#
# x=1
  u1n=(2*dx*g_L(t)+c_4*(4*u1[nx]-u1[nx-1]))/
      (2*dx*c_3+3*c_4);
  u1xx[nx]=u1n-2*u1[nx]+u1[nx-1];
#
# Interior approximation of u1xx
  for(k in 2:(nx-1)){
     u1xx[k]=u1[k+1]-2*u1[k]+u1[k-1];
  }
```

This code is explained in detail after Listing 3.2, so the details are not repeated here.

- The fractional derivative $\dfrac{\partial^\beta u_2}{\partial x^\beta}$ in eq. (6.1b) is computed according to the algorithm of eqs. (1.2).

```
#
# Approximation of u2xx
  u2xx=NULL;
#
```

```
# x=0
  u20=(2*dx*g_0(t)-c_2*(4*u2[1]-u2[2]))/
      (2*dx*c_1-3*c_2);
  u2xx_0=2*u20-5*u2[1]+4*u2[2]-u2[3];
  u2xx[1]=u2[2]-2*u2[1]+u20;
#
# x=1
  u2n=(2*dx*g_L(t)+c_4*(4*u2[nx]-u2[nx-1]))/
      (2*dx*c_3+3*c_4);
  u2xx[nx]=u2n-2*u2[nx]+u2[nx-1];
#
# Interior approximation of u2xx
  for(k in 2:(nx-1)){
    u2xx[k]=u2[k+1]-2*u2[k]+u2[k-1];
  }
```

- Eq. (6.1a) is programmed.

```
#
# u1 PDE
#
# Step through ODEs
  for(j in 1:nx){
#
#   First term in series approximation of
#   fractional derivative
    u1t[j]=A1[j,1]*u1xx_0;
#
#   Subsequent terms in series approximation
#   of fractional derivative
    for(k in 1:j){
      u1t[j]=u1t[j]+A1[j,k+1]*u1xx[k];
#
#   Next k (next term in series)
    }
    u1t[j]=cd*u1t[j]+k12*(u2[j]-u1[j]);
#
# Next j (next ODE)
  }
```

The final result is the derivative $\dfrac{\partial u_1}{\partial t}$ = u1t at the 49 interior points in x.

- Eq. (6.1b) is programmed.

```
#
# u2 PDE
#
# Step through ODEs
  for(j in 1:nx){
#
#    First term in series approximation of
#    fractional derivative
     u2t[j]=A2[j,1]*u2xx_0;
#
#    Subsequent terms in series approximation
#    of fractional derivative
     for(k in 1:j){
       u2t[j]=u2t[j]+A2[j,k+1]*u2xx[k];
#
#    Next k (next term in series)
     }
     u2t[j]=cd*u2t[j]-k12*(u2[j]-u1[j]);
#
# Next j (next ODE)
  }
```

The final result is the derivative $\dfrac{\partial u_2}{\partial t}$ = u2t at the 49 interior points in x.

- u1t,u2t are placed in ut for return to lsode.

```
#
# Place u1t, u2t in vector ut
  for(i in 1:nx){
    ut[i]   =u1t[i];
    ut[i+nx]=u2t[i];
  }
```

- The counter for the calls to pde1a is returned to the main program of Listing 6.1 via <<-.

```
#
```

```
# Increment calls to pde1a
  ncall <<- ncall+1;
```

- The ODE/MOL derivative vector is returned to lsode.

```
#
# Return derivative vector of ODEs
  return(list(c(ut)));
  }
```

list is required by lsode. c is the vector operator in R. The final } completes pde1a.

This completes the programming of eqs. (6.1), (6.2), (6.3). The output from these routines is considered next.

6.2.3 SFPDES OUTPUT

Abbreviated numerical output is given in Table 6.1.

We can observe the following details about the output in Table 6.1.

- The solution array out is $6 \times 2(49) + 1 = 99$ as explained previously.

```
[1] 6
```

```
[1] 99
```

- The order of the fractional derivatives in eqs. (6.1) as programmed in Listing 6.1, ncase=1, is confirmed.

```
    alpha = 1.50  beta = 1.50
```

- ICs (6.2) are the same since both are defined by the Gaussian function f1 (Listing 6.1).

```
    t      x     u1(x,t)    u2(x,t)
  0.00   0.00   0.00000    0.00000
  0.00   0.10   0.00000    0.00000
  0.00   0.20   0.00012    0.00012
  0.00   0.30   0.01832    0.01832
  0.00   0.40   0.36788    0.36788
  0.00   0.50   1.00000    1.00000
  0.00   0.60   0.36788    0.36788
  0.00   0.70   0.01832    0.01832
  0.00   0.80   0.00012    0.00012
  0.00   0.90   0.00000    0.00000
  0.00   1.00   0.00000    0.00000
```

Table 6.1: Abbreviated numerical output, ncase=1

```
[1] 6

[1] 99

   alpha = 1.50  beta = 1.50

     t      x    u1(x,t)    u2(x,t)
   0.00   0.00   0.00000    0.00000
   0.00   0.10   0.00000    0.00000
   0.00   0.20   0.00012    0.00012
   0.00   0.30   0.01832    0.01832
   0.00   0.40   0.36788    0.36788
   0.00   0.50   1.00000    1.00000
   0.00   0.60   0.36788    0.36788
   0.00   0.70   0.01832    0.01832
   0.00   0.80   0.00012    0.00012
   0.00   0.90   0.00000    0.00000
   0.00   1.00   0.00000    0.00000
             .              .

             .              .
   Output for t = 0.02,0.04, 0.06,
           0.08 removed

             .              .

             .              .
   0.10   0.00   0.00000    0.00000
   0.10   0.10   0.10628    0.10628
   0.10   0.20   0.19587    0.19587
   0.10   0.30   0.23641    0.23641
   0.10   0.40   0.22150    0.22150
   0.10   0.50   0.17347    0.17347
   0.10   0.60   0.11868    0.11868
   0.10   0.70   0.07140    0.07140
   0.10   0.80   0.03497    0.03497
   0.10   0.90   0.01004    0.01004
   0.10   1.00   0.00000    0.00000

   ncall = 952
```

- The solution output is for $t = 0, 0.02, ..., 0.1$ as programmed in Listing 6.1.

- The solution output is for $x = 0, 0.1, ..., 1$ as programmed in Listing 6.1 (every fifth value of x).

- The solutions for $u_1(x, t), u_2(x, t)$ remain the same as expected since the same SFPDE is programmed twice. This special case is worth considering, however, since different solutions would indicate a programming error. At $t = 0.1$,

```
0.10   0.00    0.00000    0.00000
0.10   0.10    0.10628    0.10628
0.10   0.20    0.19587    0.19587
0.10   0.30    0.23641    0.23641
0.10   0.40    0.22150    0.22150
0.10   0.50    0.17347    0.17347
0.10   0.60    0.11868    0.11868
0.10   0.70    0.07140    0.07140
0.10   0.80    0.03497    0.03497
0.10   0.90    0.01004    0.01004
0.10   1.00    0.00000    0.00000
```

The solution maximum has decreased from 1 at $t = 0$ to approximately 0.23641. The feature is demonstrated in Fig. 6.1.

- The computational effort is modest.

```
ncall = 952
```

Since the two solutions are the same, the plot of $u_2(x, t)$ is not presented.

As a second test case, if $u_2(x, t)$ is set to zero initially

```
#
# Initial condition
  nx=nx-2;
  u0=rep(0,2*nx);
  for(j in 1:nx){
    u0[j]   =f1(xj[j+1]);
    u0[j+nx]=f2(xj[j+1]);
}
```

and $k_{12} = 0$, there are no RHS terms in eq. (6.1b) that would move $u_2(x, t)$ away from zero (so it remains at zero throughout the solution). Also, eq. (6.1a) is uncoupled from eq. (6.1b),

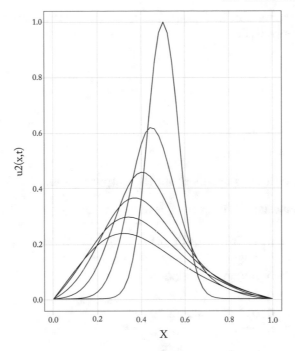

Figure 6.1: Numerical solution of eq. (6.1a), $u_1(x,t)$.

so it follows the previous response to a Gaussian. Although this case may appear trivial, it is worthwhile since if $u_1(x,t)$ and $u_2(x,t)$ do not have these expected features, a programming error would be indicated. The solution for this case is not discussed here to conserve space, but the reader can easily confirm it.

Next, several cases for $k_{12} \neq 0$ (so that eqs. (6.1a) and (6.1b) are coupled) are considered.

6.2.4   VARIATION OF THE PARAMETERS

Consideration is now given to variation of the parameters in the R routines in Listings 6.1, 6.2. As the first case, IC (6.2b) is changed to zero from the Gaussian function.

```
#
# Initial condition
  nx=nx-2;
  u0=rep(0,2*nx);
  for(j in 1:nx){
    u0[j]    =f1(xj[j+1]);
    u0[j+nx]=f2(xj[j+1]);
}
```

With k12=10 the two SFPDEs, eqs. (6.1), are connected mathematically and $u_2(x,t)$ responds to the input from $u_1(x,t)$.

```
#
# Parameters
  ncase=1;
  if(ncase==1){
    alpha=1.5;beta=1.5;k12=10;
    c_1=1;c_2=0;c_3=1;c_4=0;}
```

We can note the following details about the output in Table 6.2.

- ICs (6.2) are a Gaussian for $u_1(x, t = 0)$ and zero for $u_2(x, t = 0)$.

t	x	u1(x,t)	u2(x,t)
0.00	0.00	0.00000	0.00000
0.00	0.10	0.00000	0.00000
0.00	0.20	0.00012	0.00000
0.00	0.30	0.01832	0.00000
0.00	0.40	0.36788	0.00000
0.00	0.50	1.00000	0.00000
0.00	0.60	0.36788	0.00000
0.00	0.70	0.01832	0.00000
0.00	0.80	0.00012	0.00000
0.00	0.90	0.00000	0.00000
0.00	1.00	0.00000	0.00000

- $u_2(x, t)$ responds to $u_1(x, t)$ (through the k_{12} coupling in eqs. (6.1)).

0.02	0.00	0.00000	0.00000
0.02	0.10	0.00089	0.00018
0.02	0.20	0.02159	0.00426
0.02	0.30	0.17461	0.03446
0.02	0.40	0.46749	0.09227
0.02	0.50	0.45577	0.08996
0.02	0.60	0.21369	0.04218
0.02	0.70	0.07510	0.01482
0.02	0.80	0.02889	0.00570
0.02	0.90	0.01273	0.00251
0.02	1.00	0.00000	0.00000

- At $t = 0.1$, the two solutions are approaching each other.

Table 6.2: Abbreviated numerical output, $u_2(x, t = 0) = 0, k_{12} = 10$ (*Continues.*)

[1] 6

[1] 99

alpha = 1.50   beta = 1.50

t	x	u1(x,t)	u2(x,t)
0.00	0.00	0.00000	0.00000
0.00	0.10	0.00000	0.00000
0.00	0.20	0.00012	0.00000
0.00	0.30	0.01832	0.00000
0.00	0.40	0.36788	0.00000
0.00	0.50	1.00000	0.00000
0.00	0.60	0.36788	0.00000
0.00	0.70	0.01832	0.00000
0.00	0.80	0.00012	0.00000
0.00	0.90	0.00000	0.00000
0.00	1.00	0.00000	0.00000
0.02	0.00	0.00000	0.00000
0.02	0.10	0.00089	0.00018
0.02	0.20	0.02159	0.00426
0.02	0.30	0.17461	0.03446
0.02	0.40	0.46749	0.09227
0.02	0.50	0.45577	0.08996
0.02	0.60	0.21369	0.04218
0.02	0.70	0.07510	0.01482
0.02	0.80	0.02889	0.00570
0.02	0.90	0.01273	0.00251
0.02	1.00	0.00000	0.00000

.
.
.

Table 6.2: (*Continued.*) Abbreviated numerical output, $u_2(x, t = 0) = 0, k_{12} = 10$

```
Output for t = 0.04,0.06, 0.08,
            removed
```

```
        .                    .
        .                    .
        .                    .
0.10   0.00    0.00000    0.00000
0.10   0.10    0.06033    0.04595
0.10   0.20    0.11119    0.08468
0.10   0.30    0.13420    0.10221
0.10   0.40    0.12574    0.09576
0.10   0.50    0.09848    0.07500
0.10   0.60    0.06737    0.05131
0.10   0.70    0.04053    0.03087
0.10   0.80    0.01985    0.01512
0.10   0.90    0.00570    0.00434
0.10   1.00    0.00000    0.00000

ncall = 951
```

```
0.10   0.00    0.00000    0.00000
0.10   0.10    0.06033    0.04595
0.10   0.20    0.11119    0.08468
0.10   0.30    0.13420    0.10221
0.10   0.40    0.12574    0.09576
0.10   0.50    0.09848    0.07500
0.10   0.60    0.06737    0.05131
0.10   0.70    0.04053    0.03087
0.10   0.80    0.01985    0.01512
0.10   0.90    0.00570    0.00434
0.10   1.00    0.00000    0.00000
```

The graphical output in Figs. 6.2, 6.3 confirms these features (note the difference in the vertical scales for Figs. 6.2, 6.3).

We would expect that the solutions would continue to decrease in response to the homogeneous Dirchlet BCs (6.3), with the limiting values $u_1(x, t \to \infty) = u_2(x, t \to \infty) = 0$.

The effect of the order of the fractional derivatives, α, β, in eqs. (6.1) is demonstrated with the coding

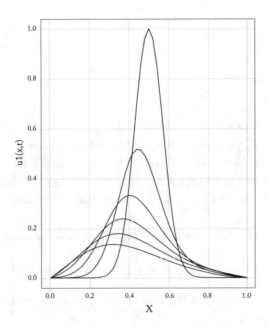

Figure 6.2: Numerical solution of eq. (6.1a), $k_{12} = 10$.

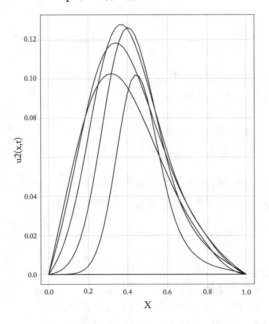

Figure 6.3: Numerical solution of eq. (6.1b), $k_{12} = 10$.

```
#
# Parameters
  ncase=1;
  if(ncase==1){
    alpha=1;beta=2;k12=10;
    c_1=1;c_2=0;c_3=1;c_4=0;}
```

Abbreviated numerical output follows in Table 6.3.

For $\alpha = 1$, (6.1a) is fully convective (hyperbolic) and for $\beta = 2$, eq. (6.1b) is fully diffusive (parabolic). The solution in Figs. 6.3, 6.4 reflect the combined effect of both of the SFPDEs (eqs. (6.1) taken simultaneously through the coupling with $k_{12} = 10$).

The graphical output is in Figs. 6.4, 6.5 (note the difference in the vertical scaling for Figs. 6.4, 6.5).

Again, we would expect the solutions to approach zero for large t in response to the homogeneous Dirichlet BCs eqs. (6.3).

The homogeneous Neumann BCs (6.4) are considered next. They are implemented with

1. Changing `c_1=1;c_2=0;c_3=1;c_4=0` (Dirichlet) to `c_1=0;c_2=1;c_3=0;c_4=1` in Listing 6.1 (with `ncase=2`).

```
#
# Parameters
  ncase=2;
  if(ncase==1){
    alpha=1.5;beta=1.5;k12=10;
    c_1=1;c_2=0;c_3=1;c_4=0;}
  if(ncase==2){
    alpha=1.5;beta=1.5;k12=10;
    c_1=0;c_2=1;c_3=0;c_4=1;}
```

2. Changing the resetting of the boundary values $u_1(x = x_l, t), u_1(x = x_u, t), u_2(x = x_l, t), u_2(x = x_u, t)$ in the main program of Listing 6.1

```
#
# Reset boundary values
  for(i in 1:nt){
    u1[i,1]=(1/3)*(4*u1[i,2]   -u1[i,3]);
    u1[i,nx]=(1/3)*(4*u1[i,nx-1]-u1[i,nx-2]);
    u2[i,1]=(1/3)*(4*u2[i,2]   -u2[i,3]);
    u2[i,nx]=(1/3)*(4*u2[i,nx-1]-u2[i,nx-2]);
    }
```

Table 6.3: Abbreviated numerical output, $\alpha = 1, \beta = 2, k_{12} = 10$

```
[1] 6

[1] 99

   alpha = 1.00   beta = 2.00

      t     x    u1(x,t)   u2(x,t)
   0.00  0.00   0.00000   0.00000
   0.00  0.10   0.00000   0.00000
   0.00  0.20   0.00012   0.00000
   0.00  0.30   0.01832   0.00000
   0.00  0.40   0.36788   0.00000
   0.00  0.50   1.00000   0.00000
   0.00  0.60   0.36788   0.00000
   0.00  0.70   0.01832   0.00000
   0.00  0.80   0.00012   0.00000
   0.00  0.90   0.00000   0.00000
   0.00  1.00   0.00000   0.00000
             .           .
             .           .
   Output for t = 0.02,0.04, 0.06,
             0.08 removed
             .           .
             .           .
   0.10  0.00   0.00000   0.00000
   0.10  0.10   0.00014   0.00004
   0.10  0.20   0.01062   0.00337
   0.10  0.30   0.17408   0.06947
   0.10  0.40   0.53180   0.31355
   0.10  0.50   0.26440   0.30484
   0.10  0.60   0.02430   0.07047
   0.10  0.70   0.00079   0.00445
   0.10  0.80   0.00001   0.00008
   0.10  0.90  -0.00000   0.00000
   0.10  1.00   0.00000   0.00000

   ncall = 624
```

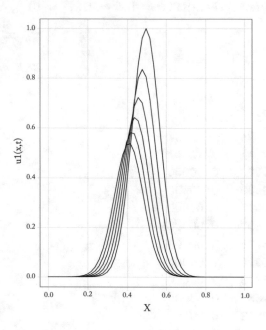

Figure 6.4: Numerical solution of eq. (6.1a), $\alpha = 1, \beta = 2, k_{12} = 10$.

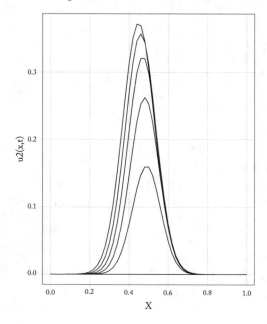

Figure 6.5: Numerical solution of eq. (6.1b), $\alpha = 1, \beta = 2, k_{12} = 10$.

The homogeneous Neumann BCs of eqs. (6.4) are implemented with non-centered FDs. For $x = x_l$ (the subscripting is now an index in x),

$$\frac{\partial u(x = x_l, t)}{\partial x} \approx \frac{-3u_1 + 4u_2 - u_3}{2\Delta x} = 0 \tag{6.6a}$$

or

$$u_1 = (1/3)(4u_2 - u_3) \tag{6.6b}$$

For $x = x_u$,

$$\frac{\partial u(x = x_u, t)}{\partial x} \approx \frac{3u_{nx} - 4u_{nx-1} + u_{nx-2}}{2\Delta x} = 0 \tag{6.6c}$$

or

$$u_{nx} = (1/3)(4u_{nx-1} - u_{nx-2}) \tag{6.6d}$$

for both $u_1(x, t)$ and $u_2(x, t)$.

3. The ODE/MOL routine of Listing 6.2 remains unchanged except that the name is changed to pde1b to keep the routines for Dirichlet and Neumann BCs distinct.

Abbreviated output follows in Table 6.4. We can note the following details about the output in Table 6.4:

- The ICs of eqs. (6.2), with $u_1(x, t = 0)$ = f1, $u_2(x, t = 0)$ = f2 are confrmed.

- The solutions for Dirichlet and Neumann BCs are not the same. Fox example,

```
Dirichlet, Table 6.2


0.10  0.00   0.00000    0.00000
0.10  0.10   0.06033    0.04595
0.10  0.20   0.11119    0.08468
0.10  0.30   0.13420    0.10221
0.10  0.40   0.12574    0.09576
0.10  0.50   0.09848    0.07500
0.10  0.60   0.06737    0.05131
0.10  0.70   0.04053    0.03087
0.10  0.80   0.01985    0.01512
0.10  0.90   0.00570    0.00434
0.10  1.00   0.00000    0.00000


Neumann,  Table 6.4


0.10  0.00   0.07952    0.06056
```

Table 6.4: Abbreviated numerical output, $\alpha = 1.5, \beta = 1.5$, Neumann BCs

```
[1] 6

[1] 99

   alpha = 1.50   beta = 1.50

     t      x    u1(x,t)   u2(x,t)
   0.00   0.00  -0.00000   0.00000
   0.00   0.10   0.00000   0.00000
   0.00   0.20   0.00012   0.00000
   0.00   0.30   0.01832   0.00000
   0.00   0.40   0.36788   0.00000
   0.00   0.50   1.00000   0.00000
   0.00   0.60   0.36788   0.00000
   0.00   0.70   0.01832   0.00000
   0.00   0.80   0.00012   0.00000
   0.00   0.90   0.00000   0.00000
   0.00   1.00  -0.00000   0.00000
          .               .
          .               .
   Output for t = 0.02,0.04, 0.06,
            0.08 removed
          .               .
          .               .
   0.10   0.00   0.07952   0.06056
   0.10   0.10   0.10720   0.08165
   0.10   0.20   0.14518   0.11057
   0.10   0.30   0.16147   0.12298
   0.10   0.40   0.14896   0.11345
   0.10   0.50   0.11916   0.09075
   0.10   0.60   0.08690   0.06619
   0.10   0.70   0.06090   0.04638
   0.10   0.80   0.04362   0.03322
   0.10   0.90   0.03445   0.02624
   0.10   1.00   0.03178   0.02421

   ncall = 950
```

0.10	0.10	0.10720	0.08165
0.10	0.20	0.14518	0.11057
0.10	0.30	0.16147	0.12298
0.10	0.40	0.14896	0.11345
0.10	0.50	0.11916	0.09075
0.10	0.60	0.08690	0.06619
0.10	0.70	0.06090	0.04638
0.10	0.80	0.04362	0.03322
0.10	0.90	0.03445	0.02624
0.10	1.00	0.03178	0.02421

In particular, u_1, u_2 at the boundaries remain at zero for the Dirichlet BCs (ncase=1) and vary with t for the Neumann BCs (ncase=2).

The solutions for Neumann are in Figs. 6.6, 6.7.

Finally, Robin BCs (6.5) are considered (with ncase=3).

```
#
# Parameters
  ncase=3;
  if(ncase==1){
    alpha=1.5;beta=1.5;k12=1;
    c_1=1;c_2=0;c_3=1;c_4=0;}
  if(ncase==2){
    alpha=1.5;beta=1.5;k12=1;
    c_1=0;c_2=1;c_3=0;c_4=1;}
  if(ncase==3){
    alpha=1.5;beta=1.5;k12=1;
    c_1=-1;c_2=1;c_3=1;c_4=1;}
```

BCs (6.5a,c) are approximated for $x = x_l$ as

$$c_2 \frac{\partial u(x = x_l, t)}{\partial x} + c_1 u(x = x_l, t) \approx c_2 \frac{-3u_1 + 4u_2 - u_3}{2\Delta x} + c_1 u_1 = 0 \qquad \text{(6.7a)}$$

or

$$u_1 = \frac{1}{3c_2/(2\Delta x) - c_1} \frac{c_2(4u_2 - u_3)}{2\Delta x} \qquad \text{(6.7b)}$$

For $x = x_u$, BCs (6.5b,d) are approximated as

$$c_4 \frac{\partial u(x = x_u, t)}{\partial x} + c_3 u(x = x_u, t) \approx c_4 \frac{3u_{nx} - 4u_{nx-1} + u_{nx-2}}{2\Delta x} + c_3 u_{nx} = 0 \qquad \text{(6.7c)}$$

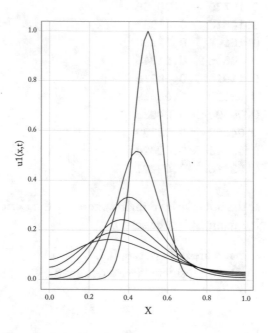

Figure 6.6: Numerical solution of eq. (6.1a), $k_{12} = 10$, Neumann BCs.

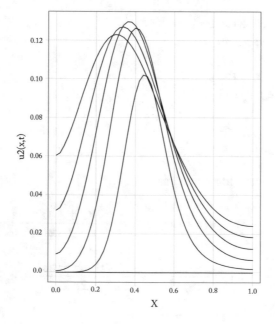

Figure 6.7: Numerical solution of eq. (6.1b), $k_{12} = 10$, Neumann BCs.

or

$$u_{nx} = \frac{1}{-3c_4/(2\Delta x) - c_3} \frac{c_4(-4u_{nx-1} + u_{nx-2})}{2\Delta x} \quad (6.7d)$$

for both $u_1(x,t)$ and $u_2(x,t)$. Note that for $c_1 = 0$, eq. (6.7b) reduces to eq. (6.6b) (Neumann BC). Also, for $c_3 = 0$, eq. (6.7d) reduces to eq. (6.6d).

Eqs. (6.7b,d) are programmed as (in Listing 6.1)

```
#
# Reset boundary values
  for(i in 1:nt){
  u1[i,1]=(1/( 3*c_2/(2*dx)-c_1))*
    c_2*( 4*u1[i,2]-u1[i,3])/(2*dx);
  u1[i,nx]=(1/(-3*c_4/(2*dx)-c_3))*
    c_4*(-4*u1[i,nx-1]+u1[i,nx-2])/(2*dx);
  u2[i,1]=(1/( 3*c_2/(2*dx)-c_1))*
    c_2*( 4*u2[i,2]-u2[i,3])/(2*dx);
  u2[i,nx]=(1/(-3*c_4/(2*dx)-c_3))*
    c_4*(-4*u2[i,nx-1]+u2[i,nx-2])/(2*dx);
  }
```

Abbreviated numerical output is given in Table 6.5.

This numerical solution is not the same as for Dirichlet and Neumann BCs, as indicated in the following comparison.

```
Dirichlet, Table 6.2


0.10  0.00   0.00000   0.00000
0.10  0.10   0.06033   0.04595
0.10  0.20   0.11119   0.08468
0.10  0.30   0.13420   0.10221
0.10  0.40   0.12574   0.09576
0.10  0.50   0.09848   0.07500
0.10  0.60   0.06737   0.05131
0.10  0.70   0.04053   0.03087
0.10  0.80   0.01985   0.01512
0.10  0.90   0.00570   0.00434
0.10  1.00   0.00000   0.00000


Neumann,  Table 6.4


0.10  0.00   0.07952   0.06056
```

Table 6.5: Abbreviated numerical output, $\alpha = 1.5, \beta = 1.5$, Robin BCs

[1] 6

[1] 99

```
alpha = 1.50  beta = 1.50

    t      x    u1(x,t)   u2(x,t)
  0.00   0.00  -0.00000   0.00000
  0.00   0.10   0.00000   0.00000
  0.00   0.20   0.00012   0.00000
  0.00   0.30   0.01832   0.00000
  0.00   0.40   0.36788   0.00000
  0.00   0.50   1.00000   0.00000
  0.00   0.60   0.36788   0.00000
  0.00   0.70   0.01832   0.00000
  0.00   0.80   0.00012   0.00000
  0.00   0.90   0.00000   0.00000
  0.00   1.00  -0.00000  -0.00000
              .             .
              .             .
              .             .
  Output for t = 0.02,0.04, 0.06,
            0.08 removed
              .             .
              .             .
              .             .
  0.10   0.00   0.07188   0.05474
  0.10   0.10   0.10305   0.07848
  0.10   0.20   0.14222   0.10831
  0.10   0.30   0.15910   0.12117
  0.10   0.40   0.14695   0.11191
  0.10   0.50   0.11737   0.08939
  0.10   0.60   0.08523   0.06491
  0.10   0.70   0.05915   0.04505
  0.10   0.80   0.04142   0.03155
  0.10   0.90   0.03115   0.02372
  0.10   1.00   0.02641   0.02011

ncall = 950
```

```
0.10   0.10    0.10720    0.08165
0.10   0.20    0.14518    0.11057
0.10   0.30    0.16147    0.12298
0.10   0.40    0.14896    0.11345
0.10   0.50    0.11916    0.09075
0.10   0.60    0.08690    0.06619
0.10   0.70    0.06090    0.04638
0.10   0.80    0.04362    0.03322
0.10   0.90    0.03445    0.02624
0.10   1.00    0.03178    0.02421


Robin, Table 6.5


0.10   0.00    0.07188    0.05474
0.10   0.10    0.10305    0.07848
0.10   0.20    0.14222    0.10831
0.10   0.30    0.15910    0.12117
0.10   0.40    0.14695    0.11191
0.10   0.50    0.11737    0.08939
0.10   0.60    0.08523    0.06491
0.10   0.70    0.05915    0.04505
0.10   0.80    0.04142    0.03155
0.10   0.90    0.03115    0.02372
0.10   1.00    0.02641    0.02011
```

The graphical output is in Fig. 6.8, 6.9.

6.3   SUMMARY AND CONCLUSIONS

The preceding example based on two simultaneous (coupled) SFPDEs (eqs. (6.1)) indicates that the MOL algorithm programming is straightforward, for Dirichlet, Neumann and Robin BCs.

Once the routines are operational, changes in the SFPDEs is straightforward. For example, nonlinear forms of the coupling terms in eqs. (6.1), such as $k_2 u_2^2 - k_1 u_1^2$ or $k_{12} u_1^a u_2^b - k_{21} u_1^c u_2^d$, requires only minor changes in the ODE/MOL routines, pde1a, pde1b, pde1c. Thus, experimentation with linear and nonlinear simultaneous SFPDEs is straightforward. Also, SFPDEs can be added by using the preceding routines as templates.

In the next chapter, SFPDEs with two-sided fractional derivative terms are considered.

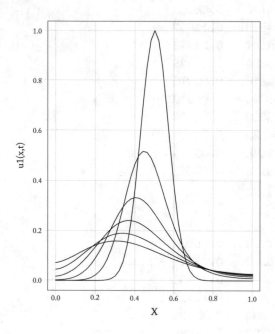

Figure 6.8: Numerical solution of eq. (6.1a), $k_{12} = 10$, Robin BCs.

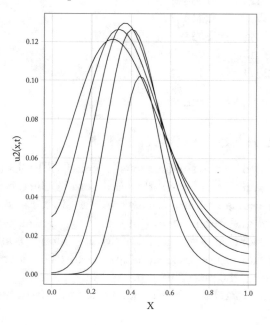

Figure 6.9: Numerical solution of eq. (6.1b), $k_{12} = 10$, Robin BCs.

CHAPTER 7

Two Sided SFPDEs

7.1 INTRODUCTION

The preceding examples of space fractional partial differential equations (SFPDEs) demonstrate the MOL solution of equations with a single fractional derivative. In this chapter we consider examples of SFPDEs which have two two-sided (two-handed) Caputo derivatives.

7.2 TWO-SIDED CONVECTIVE SFPDE, CAPUTO DERIVATIVES

The first example of a two-sided SFPDE is

$$\frac{\partial u(x,t)}{\partial t} = -v(x)\frac{\partial u(x,t)}{\partial x} + d_+\frac{\partial^\alpha u(x,t)}{\partial_+ x^\alpha} + d_-\frac{\partial^\alpha u(x,t)}{\partial_- x^\alpha} + q_s(x,t) \tag{7.1a}$$

with

$$v(x) = x; \; q_s(x,t) = \alpha e^{-t}(8x - 5x^2)$$

$$d_+ = K(x^{\alpha-1})(2-x); \; d_- = Kx(2-x)x^{\alpha-1}); \; K = \Gamma(3-\alpha)$$

and $1 < \alpha \leq 2$.

The initial condition (IC) for eq. (7.1a) is

$$u(x,t=0) = g(x) \tag{7.1b}$$

and the Dirichlet boundary conditions (BCs) are

$$u(x=a,t) = u_a(t); \; u(x=b,t) = u_b(t) \tag{7.1c,d}$$

The left-sided (+) and right-sided (−) Caputo derivatives are

$$\frac{\partial^\alpha u(x,t)}{\partial_+ x^\alpha} = \frac{1}{\Gamma(n-\alpha)}\int_a^x (x-s)^{n-\alpha-1}\frac{\partial^n u(s,t)}{\partial s^n}ds \tag{7.2a}$$

$$\frac{\partial^\alpha u(x,t)}{\partial_- x^\alpha} = \frac{(-1)^n}{\Gamma(n-\alpha)}\int_x^b (s-x)^{n-\alpha-1}\frac{\partial^n u(s,t)}{\partial s^n}ds \tag{7.2b}$$

where n is the smallest integer greater than α.[1]

   With

$$g(x) = \alpha x(2 - x); \ u_a(t) = u_b(t) = 0; \ a = 0, b = 2 \qquad (7.3a)$$

the analytical (exact) solution to eqs. (7.1) and (7.2) is[2]

$$u_a(x, t) = \alpha e^{-t} x(2 - x) \qquad (7.3b)$$

7.2.1   MAIN PROGRAM

A main program for eqs. (7.1), (7.2) and (7.3) follows.

Listing 7.1: Main program for eqs. (7.1), (7.2), (7.3)

```
#
# Two-sided SFPDE
#
# du/dt=-v(x)du/dx+d+(x,t)d^alpha u/d+x^alpha
#                 +d-(x,t)d^alpha u/d-x^alpha
#                 +qs(x,t)
#
# a < x < b, 0 < t < tf
#
# Delete previous workspaces
  rm(list=ls(all=TRUE))
#
# Access library, ODE/MOL routines
  library("deSolve");
  setwd("f:/fractional/sfpde/chap7/dirichlet");
  source("pde1a.R");
#
# Parameters
  alpha=1.7;
  K=gamma(3-alpha);
#
# Drift function
  v=function(x) x;
```

---

[1]Conventional integer derivatives are termed *local* since they apply at the spatial point where they are used. The fractional derivatives of eqs. (7.2) are *nonlocal* in the sense that they are defined in terms of integrals which are based on spatial intervals (rather than spatial points). For example, the left-sided derivative of eq. (7.2a) is defined over the spatial interval $a \leq s \leq x$. Since it applies to $x > a$, it is also designated as $+$ (x past a). Similarly, the right-sided derivative of eq. (7.2b) is defined over the spatial interval $x \leq s \leq b$. Since it applies to $x < b$, it is also designated as $-$ (x before b).

[2]The analytical solution was derived and verified with Maple and Mathematica by the first author (YS).

```
#
# Left diffusion coefficient
  d_left=function(x) K*(x^(alpha-1))*(2-x);
#
# Right diffusion coefficient
  d_right=function(x) K*x*((2-x)^(alpha-1));
#
# Source/sink function
  qs=function(x,t) alpha*exp(-t)*(8*x-5*x*x);
#
# Analytical solutnon
  ua=function(x,t) alpha*exp(-t)*x*(2-x);
#
# IC
  f=function(x,t) ua(x,0);
#
# BCs
  u_a=function(t) 0;
  u_b=function(t) 0;
#
# Grid in x
  a=0;b=2;nx=21;
  dx=(b-a)/(nx-1);
  xj=seq(from=a,to=b,by=dx);
#
# Interval in t
  t0=0;tf=1;nt=6;
  dt=(tf-t0)/(nt-1);
  tout=seq(from=0,to=tf,by=dt);
#
# a_jk coefficients
  nx=nx-1;
  A=matrix(0,nrow=nx-1,ncol=nx);
  for(j in 1:(nx-1)){
    for(k in 0:j){
      if(k==0){
      A[j,k+1]=(j-1)^(3-alpha)-j^(2-alpha)*(j-3+alpha);
      } else if(1 <= k && k<=j-1){
```

```
      A[j,k+1]=(j-k+1)^(3-alpha)-2*(j-k)^(3-alpha)+(j-k-1)^(3-
         alpha);
      } else {
      A[j,k+1]=1;
      }
    }
  }
  A=(dx^(2-alpha)/gamma(4-alpha))*A;
#
# b_jk coefficients
  B=matrix(0,nrow=nx-1,ncol=nx);
  for(j in 1:(nx-1)){
    for(k in j:nx){
      if(k==j){
        B[j,k]=1;
      } else if(j+1 <= k && k<=nx-1){
        B[j,k]=(k-j+1)^(3-alpha)-2*(k-j)^(3-alpha)+(k-j-1)^(3-
           alpha);
      } else {
        B[j,k]=(nx-j-1)^(3-alpha)-(nx-j)^(2-alpha)*(nx-j+alpha-3)
           ;
      }
    }
  }
  B=(dx^(2-alpha)/gamma(4-alpha))*B;
#
# Initial condition
  y0=rep(0,nx-1);
  for(j in 1:(nx-1)){
    y0[j]=f(xj[j+1]);
  }
  ncall=0;
#
# ODE integration
  out=lsode(y=y0,times=tout,func=pde1a,
      rtol=1e-8,atol=1e-8,maxord=5);
  nrow(out)
  ncol(out)
#
```

```
# Allocate array for u(x,t)
  nx=nx+1;
  u=matrix(0,nt,nx);
#
# u(x,t), x ne xl,xu
  for(i in 1:nt){
  for(j in 2:(nx-1)){
    u[i,j]=out[i,j];
  }
  }
#
# Reset boundary values
  for(i in 1:nt){
    u[,1] =u_a(xl,tout[i]);
    u[,nx]=u_b(xu,tout[i]);
  }
#
# Allocate array for analytical solution
  uap=matrix(0,nt,nx);
  for(i in 1:nt){
    for(j in 1:nx){
      uap[i,j]=ua((j-1)*dx,(i-1)*dt);
    }
  }
#
# Tabular numerical, analytical solutions,
# difference
  cat(sprintf("\n\n   alpha = %4.2f\n",alpha));
  cat(sprintf("\n       t      x     u(x,t)   ua(x,t)        diff"))
    ;
  for(i in 1:nt){
  iv=seq(from=1,to=nx,by=4);
  for(j in iv){
    cat(sprintf("\n %6.2f%6.2f%10.5f%10.5f%12.3e",
      tout[i],xj[j],u[i,j],uap[i,j],u[i,j]-uap[i,j]));
  }
  cat(sprintf("\n"));
  }
#
```

```
# Plot numerical, analytical solutions
  matplot(xj,t(u),type="l",lwd=2,col="black",lty=1,
    xlab="x",ylab="u(x,t)",main="");
  matpoints(xj,t(uap),pch="o",col="black");
#
# Plot error at t = tf
  err_1=abs(u[nt,]-ua(xj[1:nx],tf));
  plot(xj,err_1,type="l",xlab="x",
       ylab="Max Error at t = tf",
       main="",col="black")
#
# Calls to ODE routine
  cat(sprintf("\n    ncall = %3d\n",ncall));
```

We can note the following details about Listing 7.1.

- Brief documentation comments are followed by deletion of previous files.

```
#
# Two-sided SFPDE
#
# du/dt=-v(x)du/dx+d+(x,t)d^alpha u/d+x^alpha
#                  +d-(x,t)d^alpha u/d-x^alpha
#                  +qs(x,t)
#
# a < x < b, 0 < t < tf
#
# Delete previous workspaces
  rm(list=ls(all=TRUE))
```

- The ODE integrator library deSolve is accessed. Note that the setwd (set working directory) uses / rather than the usual \.

```
#
# Access library, ODE/MOL routines
  library("deSolve");
  setwd("f:/fractional/sfpde/chap7/dirichlet");
  source("pde1a.R");
```

- The parameters are defined. Variation of the order of the fractional derivatives, α, in eq. (7.1a) is of particular interest.

```
#
# Parameters
  alpha=1.7;
  K=gamma(3-alpha);
```

- The functions associated with eqs. (7.1) to (7.3) are defined.[3] [4]

```
#
# Drift function
  v=function(x) x;
#
# Left diffusion coefficient
  d_left=function(x) K*(x^(alpha-1))*(2-x);
#
# Right diffusion coefficient
  d_right=function(x) K*x*((2-x)^(alpha-1));
#
# Source/sink function
  qs=function(x,t) alpha*exp(-t)*(8*x-5*x*x);
```

- A function for the analytical (exact) solution of eq. (7.3b) is defined.

```
#
# Analytical solutnon
  ua=function(x,t) alpha*exp(-t)*x*(2-x);
```

- The analytical solution is used for the IC (7.1b) and BCs (7.1c,d).

```
#
# IC
  f=function(x,t) ua(x,0);
#
# BCs
  u_a=function(t) 0;
  u_b=function(t) 0;
```

- A spatial grid of 21 points is defined for the interval $x = a = 0 \le x \le x = b = 2$, so that grid values are $x = 0, 0.1, ..., 2$.

---

[3]In particular, functions for the diffusivities of eq. (7.1a), d_+, d_-, are defined.
[4]q is apparently a reserved name in R (using it produced errors). Therefore, qs was used to name the source term in eq. (7.1a).

```
#
# Grid in x
  a=0;b=2;nx=21;
  dx=(b-a)/(nx-1);
  xj=seq(from=a,to=b,by=dx);
```

- The interval in t, $t = t_0 = 0 \leq t \leq t = t_f = 1$, is defined with 6 points, so the output values are $t = 0, 0.2, ..., 1$.

```
#
# Interval in t
  t0=0;tf=1;nt=6;
  dt=(tf-t0)/(nt-1);
  tout=seq(from=0,to=tf,by=dt);
```

- The A coefficients of eq. (1.2g) are implemented as

```
#
# a_jk coefficients
  nx=nx-1;
  A=matrix(0,nrow=nx-1,ncol=nx);
  for(j in 1:(nx-1)){
    for(k in 0:j){
      if(k==0){
      A[j,k+1]=(j-1)^(3-alpha)-j^(2-alpha)*(j-3+alpha);
      } else if(1 <= k && k<=j-1){
      A[j,k+1]=(j-k+1)^(3-alpha)-2*(j-k)^(3-alpha)+(j-k-1)^(3-alpha);
      } else {
      A[j,k+1]=1;
      }
    }
  }
  A=(dx^(2-alpha)/gamma(4-alpha))*A;
```

The final factor

```
(dx^(2-alpha)/gamma(4-alpha))
```

follows from eq. (1.2j).

- The B coefficients

$$b_{j,k} = \begin{cases} 1, & k = j \\ (k-j+1)^{3-\alpha} - 2(k-j)^{3-\alpha} + (k-j-1)^{3-\alpha}, & j+1 \le k \le nx-1 \\ (nx-j-1)^{3-\alpha} - (nx-j)^{2-\alpha}(nx-j+\alpha-3), & k = nx \end{cases}$$

(7.4a)

in analogy with the A coefficients of eq. (1.2g) are implemented.

```
#
# b_jk coefficients
  B=matrix(0,nrow=nx-1,ncol=nx);
  for(j in 1:(nx-1)){
    for(k in j:nx){
      if(k==j){
        B[j,k]=1;
      } else if(j+1 <= k && k<=nx-1){
        B[j,k]=(k-j+1)^(3-alpha)-2*(k-j)^(3-alpha)+(k-j-1)^(3-alpha);
      } else {
        B[j,k]=(nx-j-1)^(3-alpha)-(nx-j)^(2-alpha)*(nx-j+alpha-3);
      }
    }
  }
  B=(dx^(2-alpha)/gamma(4-alpha))*B;
```

with the right-sided derivative of eq. (7.2b) computed as

$$I_j \approx \frac{\Delta x^{2-\alpha}}{\Gamma(4-\alpha)} \left(u_{xx,2} b_{j,nx} + \sum_{k=j}^{nx-1} u_{xx,k} b_{j,k} \right)$$

(7.4b)

in place of the left-sided derivative I_j of eq. (1.2j).

The left-sided and right-sided fractional derivatives of eqs. (7.2) are programmed via eqs. (1.2j) and (7.4b) in the ODE/MOL routine pde1a considered next.

- IC (7.1b) based on the analytical solution of eq. (7.3b) is programmed.

```
#
# Initial condition
  y0=rep(0,nx-1);
  for(j in 1:(nx-1)){
```

```
     y0[j]=f(xj[j+1]);
   }
   ncall=0;
```

The counter for the calls to pde1a is also initialzed.

- The ODE at the interior points in x are integrated by lsode.

```
#
# ODE integration
  out=lsode(y=y0,times=tout,func=pde1a,
     rtol=1e-8,atol=1e-8,maxord=5);
  nrow(out)
  ncol(out)
```

The solution matrix out has the dimensions $6 \times 21 - 2 + 1 = 20$ as demonstrated in the numerical output that follows. The offset of $+1$ indicates that the value of t is included as the first elements of each solution vector of 19 ODE solutions.

- The numerical solution is placed in matrix u (with dimensions u(6,21)).

```
#
# Allocate array for u(x,t)
  nx=nx+1;
  u=matrix(0,nt,nx);
#
# u(x,t), x ne xl,xu
  for(i in 1:nt){
  for(j in 2:(nx-1)){
    u[i,j]=out[i,j];
  }
  }
```

- The BC values, $u(x = x_l = a, t)$, $u(x = x_u = b, t)$, are reset according to the analytical solution of eq. (7.3b) (zero values).

```
#
# Reset boundary values
  for(i in 1:nt){
    u[,1] =u_a(xl,tout[i]);
    u[,nx]=u_b(xu,tout[i]);
  }
```

- The 6×21 values of the analytical solution are placed in uap.

```
#
# Allocate array for analytical solution
  uap=matrix(0,nt,nx);
  for(i in 1:nt){
    for(j in 1:nx){
      uap[i,j]=ua((j-1)*dx,(i-1)*dt);
    }
  }
```

- The numerical and analytical solutions, and their difference, are displayed.

```
#
# Tabular numerical, analytical solutions,
# difference
  cat(sprintf("\n\n    alpha = %4.2f\n",alpha));
  cat(sprintf("\n       t    x     u(x,t)    ua(x,t)          diff"));
  for(i in 1:nt){
  iv=seq(from=1,to=nx,by=4);
  for(j in iv){
    cat(sprintf("\n %6.2f%6.2f%10.5f%10.5f%12.3e",
      tout[i],xj[j],u[i,j],uap[i,j],u[i,j]-uap[i,j]));
  }
  cat(sprintf("\n"));
  }
```

Every fourth value of x is displayed from by=4.

- The numerical solution is plotted with matplot and the analytical solution is superimposed with matpoints. The two transposes, t(u), t(uap), are required so that the number of rows of u, uap equals the number of elements of xj. The solutions are plotted in t parametrically (as verified in the graphical output that follows).

```
#
# Plot numerical, analytical solutions
  matplot(xj,t(u),type="l",lwd=2,col="black",lty=1,
    xlab="x",ylab="u(x,t)",main="");
  matpoints(xj,t(uap),pch="o",col="black");
```

- The difference of the numerical and analytical solutions is plotted against x at $t = t_f = 1$.

```
#
# Plot error at t = tf
  err_1=abs(u[nt,]-ua(xj[1:nx],tf));
  plot(xj,err_1,type="l",xlab="x",
       ylab="Max Error at t = tf",
       main="",col="black")
```

- The number of calls to pde1a is displayed at the end of the solution as a measure of the required computational effort.

```
#
# Calls to ODE routine
  cat(sprintf("\n    ncall = %3d\n",ncall));
```

The ODE/MOL routine pde1a called by lsode is considered next.

7.2.2   ODE/MOL ROUTINE

A listing of pde1a follows.

Listing 7.2: ODE/MOL routine for eqs. (7.1), (7.2), (7.3)

```
  pde1a=function(t,u,parms){
#
# Function pde1a computes the derivative
# vector of the ODEs approximating the
# PDE
#
  r2dx=1/(2*dx);
  rdx2=1/dx^2;
  ut=rep(0,(nx-1));
#
# ux
  ux=NULL;
  ux_0=r2dx*(-3*u_a(t)+4*u[1]-u[2]);
  ux[1]=r2dx*(u[2]-u_a(t));
  for(k in 2:(nx-2)){
    ux[k]=r2dx*(u[k+1]-u[k-1]);
  }
  ux[nx-1]=r2dx*(u_b(t)-u[nx-2]);
  ux_nx=-r2dx*(-3*u_b(t)+4*u[nx-1]-u[nx-2]);
#
```

```
# uxx
   uxx=NULL;
   uxx_0 =rdx2*(2*u_a(t)-5*u[1]+4*u[2]-u[3]);
   uxx[1]=rdx2*(u[2] - 2*u[1]+u_a(t)); ·
   for(k in 2:(nx-2)){
     uxx[k]=rdx2*(u[k+1]-2*u[k]+u[k-1]);
   }
   uxx[nx-1]=rdx2*(u_b(t)-2*u[nx-1]+u[nx-2]);
   uxx_nx=rdx2*(2*u_b(t)-5*u[nx-1]+4*u[nx-2]-u[nx-3]);
#
# Fractional derivatives
  for(j in 1:(nx-1)){
    eq_left =0;
    eq_right=0;
    frac_diff=NULL;
    eq_left=eq_left+A[j,1]*uxx_0;
    for(k in 1:j){
      eq_left=eq_left+A[j,k+1]*uxx[k];}
    for(k in j:(nx-1)){
      eq_right=eq_right+B[j,k]*uxx[k];}
      eq_right=eq_right+B[j,nx]*uxx_nx;
#
# Sum of fractional derivatives
  frac_diff= d_left(xj[j+1])*eq_left+
             d_right(xj[j+1])*eq_right;
#
# PDE
  ut[j]=-v(xj[j+1])*ux[j]+frac_diff+qs(xj[j+1],t);
#
# Next MOL ODE
  }
#
# Increment calls to pde1a
  ncall <<- ncall+1;
#
# Return t derivative vector
  return(list(c(ut)));
}
```

We can note the following details about Listing 7.2.

- The function is defined.

```
pde1a=function(t,u,parms){
#
# Function pde1a computes the derivative
# vector of the ODEs approximating the
# PDE
```

t is the current value of t in eqs. (7.1). u is the 19-vector of ODE/MOL dependent variables. parm is an argument to pass parameters to pde1a (unused, but required in the argument list). The arguments must be listed in the order stated to properly interface with lsode called in the main program of Listing 7.1. The derivative vector of the LHS of eqs. (7.1a) is calculated next and returned to lsode.

- The factors $\dfrac{1}{2\Delta x}$, $\dfrac{1}{\Delta x^2}$ for the finite difference approximations of $\dfrac{\partial u}{\partial x}$, $\dfrac{\partial^2 u}{\partial x^2}$ are computed. Then the vector of t derivatives, ut, is defined (allocated).

```
#
  r2dx=1/(2*dx);
  rdx2=1/dx^2;
  ut=rep(0,(nx-1));
```

- The FD approximations of $\dfrac{\partial u}{\partial x}$ are computed, including the BC (7.1c,d) values, u_a(t),u_b(t).

```
#
# ux
  ux=NULL;
  ux_0=r2dx*(-3*u_a(t)+4*u[1]-u[2]);
  ux[1]=r2dx*(u[2]-u_a(t));
  for(k in 2:(nx-2)){
    ux[k]=r2dx*(u[k+1]-u[k-1]);
  }
  ux[nx-1]=r2dx*(u_b(t)-u[nx-2]);
  ux_nx=-r2dx*(-3*u_b(t)+4*u[nx-1]-u[nx-2]);
```

For example, the first derivative at the interior points is approximated as

$$\frac{\partial u}{\partial x} \approx \frac{u_{k+1} - u_{k-1}}{2\Delta x}$$

Noncentered FDs are used at the boundaries to preclude fictitious values (outside the interval in x).

- The FD approximations of $\dfrac{\partial^2 u}{\partial x^2}$ are computed, including the BC (7.1c,d) values, u_a(t),u_b(t).

```
#
# uxx
    uxx=NULL;
    uxx_0 =rdx2*(2*u_a(t)-5*u[1]+4*u[2]-u[3]);
    uxx[1]=rdx2*(u[2] - 2*u[1]+u_a(t));
    for(k in 2:(nx-2)){
      uxx[k]=rdx2*(u[k+1]-2*u[k]+u[k-1]);
    }
    uxx[nx-1]=rdx2*(u_b(t)-2*u[nx-1]+u[nx-2]);
    uxx_nx=rdx2*(2*u_b(t)-5*u[nx-1]+4*u[nx-2]-u[nx-3]);
```

For example, the second derivative at the interior points is approximated as

$$\frac{\partial^2 u}{\partial x^2} \approx \frac{u_{k+1} - 2u_k + u_{k-1}}{\Delta x^2}$$

Noncentered FDs are used at the boundaries to preclude fictitious values (outside the interval in x).

- The left-sided and right-sided fractional derivatives are computed.

```
#
# Fractional derivatives
  for(j in 1:(nx-1)){
    eq_left =0;
    eq_right=0;
    frac_diff=NULL;
    eq_left=eq_left+A[j,1]*uxx_0;
    for(k in 1:j){
      eq_left=eq_left+A[j,k+1]*uxx[k];}
    for(k in j:(nx-1)){
      eq_right=eq_right+B[j,k]*uxx[k];}
      eq_right=eq_right+B[j,nx]*uxx_nx;
```

For example, eq. (1.2j) for the left-sided derivative is implemented with

```
    eq_left =0;
    eq_left=eq_left+A[j,1]*uxx_0;
    for(k in 1:j){
    eq_left=eq_left+A[j,k+1]*uxx[k];}
```

The for implements the sum in eq. (1.2j).

Equation (7.4b) for the right-sided derivative is implemented with

```
    eq_right=0;
    for(k in j:(nx-1)){
       eq_right=eq_right+B[j,k]*uxx[k];}
       eq_right=eq_right+B[j,nx]*uxx_nx;
```

The for implements the sum in eq. (7.4b).

- The two fractional derivatives are summed according to eq. (7.1a).

```
#
# Sum of fractional derivatives
   frac_diff= d_left(xj[j+1])*eq_left+
            d_right(xj[j+1])*eq_right;
```

for(j in 1:(nx-1)){ steps through the 19 ODEs at the interior points in x.

- The MOL approximation of eq. (7.1a) at point j in x is programmed as

```
#
# PDE
   ut[j]=-v(xj[j+1])*ux[j]+frac_diff+qs(xj[j+1],t);
#
# Next MOL ODE
   }
```

The for in j is then continued.

- The counter for the calls to pde1s is incremented and returned to the main program of Listing 7.1 with <<-.

```
#
# Increment calls to pde1a
   ncall <<- ncall+1;
```

• The t derivative vector is returned to lsode as a list (as required by lsode).

```
#
# Return t derivative vector
  return(list(c(ut)));
}
```

The final } concludes pde1a.

The numerical and graphical output from the preceding routines is next.

7.2.3   SFPDE OUTPUT

Abbreviated numerical output from the routines of Listings 7.1 and 7.2 is shown in Table 7.1. We can note the following details about the output in Table 7.1.

• The solution array out is $6 \times 19 + 1 = 20$.

```
[1] 6
```

```
[1] 20
```

The offset of +1 indicates that t is also included as the first element of the 6 solution vectors.

• The order of the fractional derivatives in eq. (7.1a) is displayed (and can be varied).

```
alpha = 1.70
```

• IC (7.1b) is the same for both the numerical and analytical solutions since both are defined by the analytical solution of eq. (7.3b) with $t = 0$.

• The solution output is for $t = 0, 0.2, ..., 1$ as programmed in Listing 7.1.

• The solution output is for $x = 0, 0.4, ..., 2$ as programmed in Listing 7.1 (every fourth value of x).

• The maximum error in Table 7.1 is -1.063e-08.

• The computational effort is modest, ncall = 222.

The agreement between the numerical and analytical solutions is confirmed in Figs. 7.1, 7.2.

This concludes the two-sided, SFPDE with Caputo derivative. We now consider an analogous example with Riemann-Liouville derivatives.

Table 7.1: Abbreviated numerical output for eqs. (7.1), (7.2), (7.3)

```
[1] 6

[1] 20

    alpha = 1.70

       t      x      u(x,t)     ua(x,t)          diff
     0.00   0.00    0.00000     0.00000      0.000e+00
     0.00   0.40    1.08800     1.08800      0.000e+00
     0.00   0.80    1.63200     1.63200      0.000e+00
     0.00   1.20    1.63200     1.63200      0.000e+00
     0.00   1.60    1.08800     1.08800      0.000e+00
     0.00   2.00    0.00000     0.00000      0.000e+00


     0.20   0.00    0.00000     0.00000      0.000e+00
     0.20   0.40    0.89078     0.89078     -6.536e-09
     0.20   0.80    1.33617     1.33617     -1.009e-08
     0.20   1.20    1.33617     1.33617     -1.063e-08
     0.20   1.60    0.89078     0.89078     -8.027e-09
     0.20   2.00    0.00000     0.00000      0.000e+00
                 .                       .
                 .                       .
                 .                       .

     Output for t = 0.4, 0.6, 0.8 removed

                 .                       .
                 .                       .
                 .                       .

     1.00   0.00    0.00000     0.00000      0.000e+00
     1.00   0.40    0.40025     0.40025      4.228e-09
     1.00   0.80    0.60038     0.60038      6.514e-09
     1.00   1.20    0.60038     0.60038      6.841e-09
     1.00   1.60    0.40025     0.40025      5.135e-09
     1.00   2.00    0.00000     0.00000      0.000e+00


    ncall = 222
```

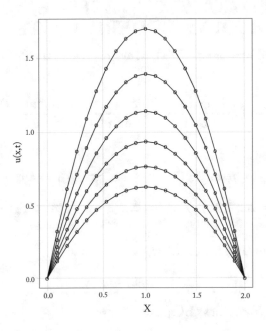

Figure 7.1: Numerical, analytical solutions of eqs. (7.1), (7.2), (7.3).

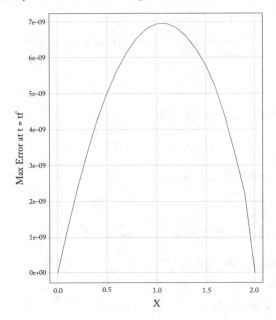

Figure 7.2: Error in the numerical solution of eqs. (7.1), (7.2), (7.3).

7.3   TWO-SIDED CONVECTIVE SFPDE, RIEMANN-LIOUVILLE DERIVATIVES

Equation (7.1a) is restated here as eq. (7.5a), with a change in the diffusivities, source terms and BCs.

$$\frac{\partial u(x,t)}{\partial t} = -v(x)\frac{\partial u(x,t)}{\partial x} + d_+ \frac{\partial^\alpha u(x,t)}{\partial_+ x^\alpha} + d_- \frac{\partial^\alpha u(x,t)}{\partial_- x^\alpha} + q_s(x,t) \tag{7.5a}$$

with

$$v(x) = x; \; q_s(x,t) = q_{s1}\alpha e^{-t}x(2-x) - \alpha e^{-t}x^2; \; q_{s1} = 4(\alpha-1)(\alpha-3)(\alpha-4)$$

$$d_+ = Kx^\alpha(2-x); \; d_- = Kx(2-x)^\alpha; \; K = \Gamma(5-\alpha)$$

and $1 < \alpha \le 2$.

The initial condition (IC) for eq. (7.5a) is

$$u(x,t=0) = g(x) \tag{7.5b}$$

and the Robin boundary conditions (BCs) are

$$e^t \frac{\partial u(x=a,t)}{\partial x} + u(x=a,t) = 2\alpha \tag{7.5c}$$

$$-\frac{1}{2\alpha}\frac{\partial u(x=b,t)}{\partial x} + u(x=b,t) = e^{-t} \tag{7.5d}$$

$g(x)$ in IC (7.5b) is

$$g(x) = \alpha x(2-x) \tag{7.6a}$$

The analytical (exact) solution to eqs. (7.5) is

$$u_a(x,t) = \alpha e^{-t}x(2-x) \tag{7.6b}$$

The routines for eqs. (7.5) and (7.6) follow.

7.3.1   MAIN PROGRAM

A main program for eqs. (7.5) and (7.6) is listed next.

Listing 7.3: Main program for eqs. (7.5), (7.6)

```
#
# SFPDE, Riemann-Liouville derivative
#
# du/dt=-v(x)du/dx+d+(x,t)d^alpha u/d+x^alpha
#                 +d-(x,t)d^alpha u/d-x^alpha
```

```
#                      +qs(x,t)
#
# a < x < b, 0 < t < tf
#
# Delete previous workspaces
  rm(list=ls(all=TRUE))
#
# Access library, ODE/MOL routines
  library("deSolve");
  source("f:/fractional/sfpde/chap7/robin/pde1a.R");
#
# Parameters
  alpha=1.5;
  K=gamma(5-alpha);
  qs1=4*(alpha-1)*(alpha-3)*(alpha-4);
#
# Drift function
  v=function(x,t) x;
#
# Left diffusion coefficient
  d_left=function(x,t) K*(x^alpha)*(2-x);
#
# Right diffusion coefficient
  d_right=function(x,t) K*x*((2-x)^alpha);
#
# Source/sink function
  qs=function(x,t) qs1*alpha*exp(-t)*x*(2-x)-
                   alpha*exp(-t)*x^2;
#
# Analytical solution
  ua=function(x,t) alpha*exp(-t)*x*(2-x);
#
# IC
  f=function(x) ua(x,0);
#
# BC functions
  g_0=function(t) 2*alpha;
  g_L=function(t) exp(-t);
#
```

```
# BC coefficients
  c_1=function(t) 1;
  c_2=function(t) exp(t);
  c_3=function(t) 1;
  c_4=function(t) -1/(2*alpha);
#
# Grid in x
  a=0;b=2;nx=21;
  dx=(b-a)/(nx-1);
  xj=seq(from=a,to=b,by=dx);
#
# Interval in t
  t0=0;tf=1;nt=6;
  dt=(tf-t0)/(nt-1);
  tout=seq(from=0,to=tf,by=dt);
#
# a_jk coefficients
  nx=nx-1;
  A=matrix(0,nrow=nx-1,ncol=nx);
  for(j in 1:(nx-1)){
    for(k in 0:j){
    if (k == 0) {
      A[j,k+1]=(j-1)^(3-alpha)-j^(2-alpha)*(j-3+alpha);
    } else if (1 <= k && k<=j-1) {
      A[j,k+1]=(j-k+1)^(3-alpha)-2*(j-k)^(3-alpha)+(j-k-1)^(3-
          alpha);
    } else
      A[j,k+1] = 1;
    }
  }
  A=(dx^(2-alpha)/gamma(4-alpha))*A;
#
# b_jk coefficients
  B=matrix(0,nrow=nx-1,ncol=nx);
  for(j in 1:(nx-1)){
    for(k in j:nx){
    if (k == j) {
      B[j,k] = 1;
    } else if (j+1 <= k && k<=nx-1) {
```

```
    B[j,k]=(k-j+1)^(3-alpha)-2*(k-j)^(3-alpha)+(k-j-1)^(3-alpha
        );
  } else
    B[j,k]=(nx-j-1)^(3-alpha)-(nx-j)^(2-alpha)*(nx-j+alpha-3);
  }
}
B=(dx^(2-alpha)/gamma(4-alpha))*B;
#
# ICs
  u0=rep(0,nx-1);
  for(j in 1:(nx-1)){
    u0[j]=f(xj[j+1]);}
  ncall=0;
#
# ODE integration
  out=lsode(y=u0,times=tout,func=pde1a,
            rtol=1e-6,atol=1e-6,maxord=5);
  nrow(out)
  ncol(out)
#
# Array for u(x,t)
  nx=nx+1;
  u=matrix(0,nt,nx);
#
# Reset boundary values
  for(i in 1:nt){
    u[i,1]=(2*dx*g_0(tout[i])-
          c_2(tout[i])*(4*out[i,2]-out[i,3]))/
          (2*dx*c_1(tout[i])-3*c_2(tout[i]));}
  for(i in 1:nt){
    u[i,nx]=(-2*dx*g_L(tout[i])-
          c_4(tout[i])*(4*out[i,nx-1]-out[i,nx-2]))/
          (-2*dx*c_3(tout[i])-3*c_4(tout[i]));}
#
# u(x,t) at interior points
  for(i in 1:nt){
  for(j in 2:(nx-1)){
    u[i,j]=out[i,j];
  }
```

```
    }
  #
  # Analytical solution
    uap=matrix(0,nt,nx);
    for(i in 1:nt){
      for(j in 1:nx){
        uap[i,j]=ua((j-1)*dx,(i-1)*dt);
      }
    }
  #
  # Tabular numerical, analytical solutions,
  # difference
    cat(sprintf("\n\n    alpha = %4.2f\n",alpha));
    cat(sprintf("\n        t       x     u(x,t)    ua(x,t)           diff"))
        ;
    for(i in 1:nt){
    iv=seq(from=1,to=nx,by=4);
    for(j in iv){
      cat(sprintf("\n %6.2f%6.2f%10.5f%10.5f%12.3e",
        tout[i],xj[j],u[i,j],uap[i,j],u[i,j]-uap[i,j]));
    }
    cat(sprintf("\n"));
    }
  #
  # Plot numerical, analytical solutions
    matplot(xj,t(u),type="l",lwd=2,col="black",lty=1,
      xlab="x",ylab="u(x,t)",main="");
    matpoints(xj,t(uap),pch="o",col="black");
  #
  # Plot error at t = tf
    err_1=abs(u[nt,]-ua(xj[1:nx],tf));
    plot(xj,err_1,type="l",xlab="x",
        ylab="Max Error at t = tf",
        main="",col="black")
  #
  # Calls to ODE routine
    cat(sprintf("\n    ncall = %3d\n",ncall));
```

Listing 7.3 is similar to Listing 7.1, so only the differences are indicated here.

- The ODE/MOL routine is again named pde1a, but it is a different routine than in Listing 7.2 (and discussed subsequently).

```
#
# Access library, ODE/MOL routines
  library("deSolve");
  source("f:/fractional/sfpde/chap7/robin/pde1a.R");
```

- The parameters are specified.

```
#
# Parameters
  alpha=1.5;
  K=gamma(5-alpha);
  qs1=4*(alpha-1)*(alpha-3)*(alpha-4);
```

- The functions pertaining to eqs. (7.5a), (7.6) are defined.

```
#
# Drift function
  v=function(x,t) x;
#
# Left diffusion coefficient
  d_left=function(x,t) K*(x^alpha)*(2-x);
#
# Right diffusion coefficient
  d_right=function(x,t) K*x*((2-x)^alpha);
#
# Source/sink function
  qs=function(x,t) qs1*alpha*exp(-t)*x*(2-x)-
                   alpha*exp(-t)*x^2;
#
# Analytical solution
  ua=function(x,t) alpha*exp(-t)*x*(2-x);
```

- The function for IC (7.5b) is defined in terms of the analytical solution of eq. (7.6b).

```
#
# IC
  f=function(x) ua(x,0);
```

- The coefficients and RHS functions for BCs (7.5c,d) are defined.

```
#
# BC functions
  g_0=function(t) 2*alpha;
  g_L=function(t) exp(-t);
#
# BC coefficients
  c_1=function(t) 1;
  c_2=function(t) exp(t);
  c_3=function(t) 1;
  c_4=function(t) -1/(2*alpha);
```

- A spatial grid of 21 points is defined for the interval $x = a = 0 \leq x \leq x = b = 2$, so that grid values are $x = 0, 0.1, ..., 2$

- The interval in t, $t = t_0 = 0 \leq t \leq t = t_f = 1$, is defined with 6 points, so the output values are $t = 0, 0.2, ..., 1$.

```
#
# Interval in t
  t0=0;tf=1;nt=6;
  dt=(tf-t0)/(nt-1);
  tout=seq(from=0,to=tf,by=dt);
```

- The 19 ODEs at the interior points in x are integrated with lsode.

```
#
# ODE integration
  out=lsode(y=u0,times=tout,func=pde1a,
            rtol=1e-6,atol=1e-6,maxord=5);
  nrow(out)
  ncol(out)
```

- The numerical solution is placed in array u. This includes the use of BCs (7.5c,d), with code adapted from pde1a considered next and explained with eq. (3.5c).

```
#
# Array for u(x,t)
  nx=nx+1;
  u=matrix(0,nt,nx);
```

```
#
# Reset boundary values
  for(i in 1:nt){
    u[i,1]=(2*dx*g_0(tout[i])-
           c_2(tout[i])*(4*out[i,2]-out[i,3]))/
           (2*dx*c_1(tout[i])-3*c_2(tout[i]));}
  for(i in 1:nt){
    u[i,nx]=(-2*dx*g_L(tout[i])-
            c_4(tout[i])*(4*out[i,nx-1]-out[i,nx-2]))/
            (-2*dx*c_3(tout[i])-3*c_4(tout[i]));}
#
# u(x,t) at interior points
  for(i in 1:nt){
  for(j in 2:(nx-1)){
    u[i,j]=out[i,j];
  }
  }
```

- The coding for the tabular and graphical output of the numerical and analytical solutions is the same as in Listing 7.1

The MOL/ODE routine pde1a is considered next.

7.3.2   ODE/MOL ROUTINE

pde1a is listed below.

Listing 7.4: ODE/MOL for eqs. (7.5), (7.6)

```
  pde1a=function(t,u,parms){
#
# Function pde1a computes the derivative
# vector of the ODEs approximating the
# PDE
#
# Array for t derivative vector
  ut=rep(0,nx-1);
#
# ux
  ux=NULL;
  u0=(2*dx*g_0(t)-
     c_2(t)*(4*u[1]-u[2]))/(2*dx*c_1(t)-3*c_2(t));
```

```
  un=(-2*dx*g_L(t)-
      c_4(t)*(4*u[nx-1]-u[nx-2]))/(-2*dx*c_3(t)-3*c_4(t));
  ux_0=(1/(2*dx))*(-3*u0+4*u[1]-u[2]);
  ux[1]=(1/(2*dx))*(u[2]-u0);
  for(k in 2:(nx-2)){
    ux[k]=(1/(2*dx))*(u[k+1]-u[k-1]);}
  ux[nx-1]=(1/(2*dx))*(un-u[nx-2]);
  ux_n    =(-1/(2*dx))*(-3*un+4*u[nx-1]-u[nx-2]);
#
# uxx
  uxx=NULL;
  uxx_0=(1/(dx^2))*(2*u0-5*u[1]+4*u[2]-u[3]);
  uxx[1]=(1/(dx^2))  *(u[2]-2*u[1]+u0);
  for(k in 2:(nx-2)){
    uxx[k]=(1/(dx^2))*(u[k+1]-2*u[k]+u[k-1]);}
  uxx[nx-1]=(1/(dx^2))*(un-2*u[nx-1]+u[nx-2]);
  uxx_n    =(1/(dx^2))*(2*un-5*u[nx-1]+4*u[nx-2]-u[nx-3]);
#
# Fractional derivatives
  for(j in 1:(nx-1)){
    eq_left=((xj[j+1]-a)^(-alpha)*u0)/gamma(1-alpha)+
            ((xj[j+1]-a)^(1-alpha)*ux_0)/gamma(2-alpha);
   eq_right=((b-xj[j+1])^(-alpha)*un)/gamma(1-alpha)-
            ((b-xj[j+1])^(1-alpha)*ux_n)/gamma(2-alpha);
#
#   Sum of fractional derivatives
    frac_diff=NULL;
    eq_left=eq_left+A[j,1]*uxx_0;
    for(k in 1:j){
      eq_left=eq_left+A[j,k+1]*uxx[k];}
      for(k in j:(nx-1)){
        eq_right=eq_right+B[j,k]*uxx[k];}
        eq_right=eq_right+B[j,nx]*uxx_n;
    frac_diff=d_left(xj[j+1],t)*eq_left+
              d_right(xj[j+1],t)*eq_right;
#
# PDE
    ut[j]=-v(xj[j+1],t)*ux[j]+frac_diff+qs(xj[j+1],t);
#
```

```
# Next ODE
  }
#
# Increment call to pde1a
  ncall <<- ncall+1;
#
# Return t derivative vector
  return(list(c(ut)));
  }
```

We can note the following details about Listing 7.4.

- The function is defined

```
pde1a=function(t,u,parms){
#
# Function pde1a computes the derivative
# vector of the ODEs approximating the
# PDE
#
# Array for t derivative vector
  ut=rep(0,nx-1);
```

The input arguments are discussed after Listing 7.2. ut for the t derivative vector is defined.

- $\dfrac{\partial u}{\partial x}$ is computed, starting with $u(x = a, t)$= u0, $u(x = b, t)$= un according to eq. (3.5c).

```
#
# ux
  ux=NULL;
  u0=(2*dx*g_0(t)-
      c_2(t)*(4*u[1]-u[2]))/(2*dx*c_1(t)-3*c_2(t));
  un=(-2*dx*g_L(t)-
      c_4(t)*(4*u[nx-1]-u[nx-2]))/(-2*dx*c_3(t)-3*c_4(t));
```

- The FD approximation

$$\frac{\partial u(x = a, t)}{\partial x} \approx \frac{-3u(x = a, t) + 4u(x = a + \Delta x) - u(x = a + 2\Delta x, t)}{2\Delta x}$$

is coded as

```
ux_0=(1/(2*dx))*(-3*u0+4*u[1]-u[2]);
```

- The FD approximation

$$\frac{\partial u(x = a + \Delta x, t)}{\partial x} \approx \frac{u(x = a + 2\Delta x, t) - u(x = a, t)}{2\Delta x}$$

is coded as

```
ux[1]=(1/(2*dx))*(u[2]-u0);
```

- The FD approximation

$$\frac{\partial u(x, t)}{\partial x} \approx \frac{u(x + \Delta x, t) - u(x - \Delta x, t)}{2\Delta x}$$

is coded as

```
for(k in 2:(nx-2)){
  ux[k]=(1/(2*dx))*(u[k+1]-u[k-1]);}
```

for $x = a + k\Delta x, k = 2, 3, ..., nx - 2$

- The FD approximation

$$\frac{\partial u(x = b - \Delta x, t)}{\partial x} \approx \frac{u(x = b, t) - u(x = b - 2\Delta x, t)}{2\Delta x}$$

is coded as

```
ux[nx-1]=(1/(2*dx))*(un-u[nx-2]);
```

- The FD approximation

$$\frac{\partial u(x = b, t)}{\partial x} \approx \frac{3u(x = b, t) - 4u(x = b - \Delta x) + u(x = b - 2\Delta x, t)}{2\Delta x}$$

is coded as

```
ux_n=(-1/(2*dx))*(-3*un+4*u[nx-1]-u[nx-2]);
```

- The programming for the second derivative $\frac{\partial^2 u}{\partial x^2}$ is similar.

```
#
# uxx
  uxx=NULL;
  uxx_0=(1/(dx^2))*(2*u0-5*u[1]+4*u[2]-u[3]);
  uxx[1]=(1/(dx^2)) *(u[2]-2*u[1]+u0);
  for(k in 2:(nx-2)){
    uxx[k]=(1/(dx^2))*(u[k+1]-2*u[k]+u[k-1]);}
  uxx[nx-1]=(1/(dx^2))*(un-2*u[nx-1]+u[nx-2]);
  uxx_n    =(1/(dx^2))*(2*un-5*u[nx-1]+4*u[nx-2]-u[nx-3]);
```

FD approximations for the second derivative, from eqs. (1.2h,i), are used in place of the FD approximations for the first derivative. For example,

$$\frac{\partial u^2(x,t)}{\partial x^2} \approx \frac{u(x+\Delta x,t) - 2u(x,t) + u(x-\Delta x,t)}{\Delta x^2}$$

is coded as

```
  for(k in 2:(nx-2)){
    uxx[k]=(1/(dx^2))*(u[k+1]-2*u[k]+u[k-1]);}
```

for $x = a + k\Delta x, k = 2, 3, ..., nx - 2$

• The fractional derivatives in eq. (7.5a) are of the Riemann-Liouville form (rather than the Caputo form of eqs. (7.2)).

$$\frac{RL\partial^\alpha u(x,t)}{\partial_+ x^\alpha} = \frac{1}{\Gamma(n-\alpha)} \frac{\partial^n}{\partial x^n} \int_a^x (x-s)^{n-\alpha-1} u(s,t)ds \tag{7.7a}$$

$$\frac{RL\partial^\alpha u(x,t)}{\partial_- x^\alpha} = \frac{(-1)^n}{\Gamma(n-\alpha)} \frac{\partial^n}{\partial x^n} \int_x^b (s-x)^{n-\alpha-1} u(s,t)ds \tag{7.7b}$$

where n is again the smallest integer greater than α. RL is added at this point to designate a Riemann-Liouville fractional derivative. A superscript C for Caputo is not used. Rather, it is assumed that the fractional derivatives are Caputo unless designated otherwise (with RL).

In order to use the preceding algorithm and coding for a Caputo derivative, the conversion between the two types of derivatives is used. For $1 < \alpha < 2$ ([1], eq. (2.20)),

$$\frac{RL\partial^\alpha u(x,t)}{\partial_- x^\alpha} = \frac{(x-a)^{-\alpha} u(x=a,t)}{\Gamma(1-\alpha)} + \frac{(x-a)^{1-\alpha} \partial u(x=a,t)/\partial x}{\Gamma(2-\alpha)} + \frac{\partial^\alpha u(x,t)}{\partial_- x^\alpha}$$

$$\tag{7.8a}$$

$$\frac{{}^{RL}\partial^{\alpha}u(x,t)}{\partial_{+}x^{\alpha}} = \frac{(b-x)^{-\alpha}u(x=b,t)}{\Gamma(1-\alpha)} + \frac{(-x)^{1-\alpha}\partial u(x=b,t)/\partial x}{\Gamma(2-\alpha)} + \frac{\partial^{\alpha}u(x,t)}{\partial_{+}x^{\alpha}} \quad (7.8b)$$

The coding for the conversion constants

$$\frac{(x-a)^{-\alpha}u(x=a,t)}{\Gamma(1-\alpha)} + \frac{(x-a)^{1-\alpha}\partial u(x=a,t)/\partial x}{\Gamma(2-\alpha)} \quad (7.8c)$$

$$\frac{(b-x)^{-\alpha}u(x=b,t)}{\Gamma(1-\alpha)} + \frac{(-x)^{1-\alpha}\partial u(x=b,t)/\partial x}{\Gamma(2-\alpha)} \quad (7.8d)$$

is

```
#
# Fractional derivatives
  for(j in 1:(nx-1)){
    eq_left=((xj[j+1]-a)^(-alpha)*u0)/gamma(1-alpha)+
            ((xj[j+1]-a)^(1-alpha)*ux_0)/gamma(2-alpha);
    eq_right=((b-xj[j+1])^(-alpha)*un)/gamma(1-alpha)-
            ((b-xj[j+1])^(1-alpha)*ux_n)/gamma(2-alpha);
```

- Equation (7.5a) is programmed as

```
#
# PDE
    ut[j]=-v(xj[j+1],t)*ux[j]+frac_diff+qs(xj[j+1],t);
#
# Next ODE
  }
```

This provides the vector of ODE/MOL t derivatives (ut) which is then integrated by lsode for the next step along the solution.

- The counter for the calls to pde1a is incremented and returned to the main program of Listing 7.3 with <<-.

```
#
# Increment call to pde1a
  ncall <<- ncall+1;
```

- The derivative vector ut is returned to lsode as a list.

```
#
# Return t derivative vector
  return(list(c(ut)));
  }
```

The final } concludes pde1a.

The output from the routines in Listings 7.3, 7.4 is considered next.

7.3.3   SFPDE OUTPUT

Abbreviated numerical output for eqs. (7.5), (7.6) is shown in Table 7.2.
      We can note the following details about the output in Table 7.2.

- The solution array out is $6 \times 19 + 1 = 20$.

    ```
    [1] 6
    ```

    ```
    [1] 20
    ```

    The offset of +1 indicates that t is also included as the first element of the 6 solution vectors.

- The order of the fractional derivatives in eq. (7.1a) is displayed (and can be varied).

    ```
    alpha = 1.50
    ```

- IC (7.5b) is the same for both the numerical and analytical solutions since both are defined by the analytical solution of eq. (7.6b) with $t = 0$.

- The solution output is for $t = 0, 0.2, ..., 1$ as programmed in Listing 7.3.

- The solution output is for $x = 0, 0.4, ..., 2$ as programmed in Listing 7.3 (every fourth value of x).

- The maximum error in Table 7.1 is $-4.727e-06$.

- The computational effort is modest, ncall = 126.

The agreement between the numerical and analytical solutions is confirmed in Figs. 7.3, 7.4. This concludes the two-sided, SFPDE with RL derivatives.

Table 7.2: Abbreviated numerical output for eqs. (7.5), (7.6)

```
[1] 6

[1] 20

  alpha = 1.50

    t     x     u(x,t)   ua(x,t)      diff
  0.00  0.00   0.00000   0.00000   0.000e+00
  0.00  0.40   0.96000   0.96000   0.000e+00
  0.00  0.80   1.44000   1.44000   0.000e+00
  0.00  1.20   1.44000   1.44000   0.000e+00
  0.00  1.60   0.96000   0.96000   0.000e+00
  0.00  2.00   0.00000   0.00000   0.000e+00


  0.20  0.00  -0.00000   0.00000  -4.884e-07
  0.20  0.40   0.78598   0.78598  -2.799e-07
  0.20  0.80   1.17897   1.17897  -1.929e-07
  0.20  1.20   1.17897   1.17897  -2.502e-07
  0.20  1.60   0.78598   0.78598  -4.517e-07
  0.20  2.00  -0.00000   0.00000  -9.899e-07
             .                    .
             .                    .
             .                    .

      Output for t = 0.4, 0.6, 0.8 removed
             .                    .
             .                    .
             .                    .
  1.00  0.00  -0.00000   0.00000  -2.139e-06
  1.00  0.40   0.35316   0.35316  -2.092e-06
  1.00  0.80   0.52974   0.52975  -2.579e-06
  1.00  1.20   0.52974   0.52975  -2.965e-06
  1.00  1.60   0.35316   0.35316  -3.190e-06
  1.00  2.00  -0.00000   0.00000  -4.727e-06


  ncall = 126
```

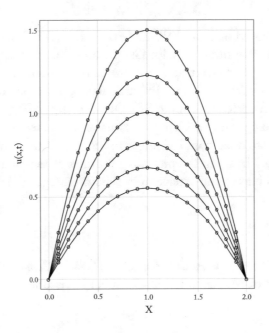

Figure 7.3: Numerical, analytical solutions of eqs. (7.5), (7.6).

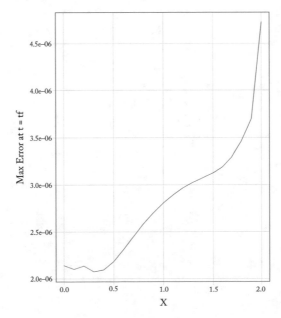

Figure 7.4: Error in the numerical solution of eqs. (7.5), (7.6).

7.4   SUMMARY AND CONCLUSIONS

The preceding examples based on two SFPDEs (eqs. (7.1a), (7.5a)) indicates that the MOL algorithm programming is straightforward, for Caputo and Riemann-Liouville derivatives. Also, a methodology is detailed for converting a test problem with one type of fractional derivative to the other, as illustrated with eq. (7.5a). In particular, applications of RL SFPDEs reported in the literature can be converted to Caputo SFPDEs thereby adding a variety of test problems (through the numerical algorithm for Caputo derivatives discussed in the preceding chapters).

Once the routines are operational, changes in the SFPDEs are straightforward. Alternative SFPDEs can be added by using the preceding routines as templates.

In the next (final) chapter, the extension of classical integer PDEs to SFPDEs is considered.

REFERENCES

[1] Ishteva, M.K. (2005), *Properies and Applications of the Caputo Fractional Operator*, M.S. Thesis, Department of Mathematics, Universitaet Karlsruhe. 249

CHAPTER 8

Integer to Fractional Extensions

8.1   INTRODUCTION

In this concluding chapter, we consider additional space fractional partial differential equations (SFPDEs) by extending some classical (legacy) integer space PDEs (ISPDEs) to SFPDEs. The ISPDEs have analytical (exact) solutions that can be used to test the SFPDE routines as a special case for which the fractional derivative order is integer.

8.2   FRACTIONAL DIFFUSION EQUATION

The first example application is the fractional diffusion equation with Dirichlet, Neumann and Robin boundary conditions (BCs).

$$\frac{\partial u(x,t)}{\partial t} = \frac{\partial^\alpha u(x,t)}{\partial x^\alpha} \tag{8.1a}$$

with the order of the fractional derivative $1 \le \alpha \le 2$. For $\alpha = 2$, the analytical solution is used to test the fractional order routines.

The initial condition (IC) for eq. (8.1a) is

$$u(x, t = 0) = u_a(x, t = 0) \tag{8.1b}$$

and the Dirichlet boundary conditions (BCs) are

$$u(x = x_l, t) = u(x = x_u, t) = 0 \tag{8.1c,d}$$

The integer exact solution is

$$u_a(x,t) = e^{-\pi^2 t} \sin(\pi x) \tag{8.1e}$$

The routines for the method of lines (MOL) solution of eqs. (8.1) are next.

8.2.1   MAIN PROGRAM, DIRCHLET BCS

A main program for eqs. (8.1) follows.

Listing 8.1: Main program for eqs. (8.1).

```
#
# Diffusion equation
#
#    ut=D*(d^alpha u/dx^alpha)
#
#    xl < x < xu, 0 < t < tf, xl=0, xu=1, D=1
#
#    u(x,t=0)=sin(pi*x)
#
#    u(x=xl,t)=u(x=xu,t)=0
#
#    ua(x,t)=exp(-pi^2*t)*sin(pi*x) (alpha=2)
#
# Delete previous workspaces
  rm(list=ls(all=TRUE))
#
# Access functions for numerical solution
  library("deSolve");
  setwd("f:/fractional/sfpde/chap8/ex1/dirichlet");
  source("pde1a.R");
#
# Parameters
  ncase=1;
  if(ncase==1){alpha=2};
  if(ncase==2){alpha=1.5};
  D=1;
#
# Analytical solution
  ua=function(x,t) exp(-pi^2*t)*sin(pi*x);
#
# Initial condition function (IC)
  f=function(x) sin(pi*x);
#
# Boundary condition functions (BCs)
  g_0=function(t) 0;
  g_L=function(t) 0;
#
# Boundary condition coefficients
```

```
  c_1=function(t) 1;
  c_2=function(t) 0;
  c_3=function(t) 1;
  c_4=function(t) 0;
#
# Spatial grid
  xl=0;xu=1;nx=41;dx=(xu-xl)/(nx-1);
  xj=seq(from=xl,to=xu,by=dx);
  cd=dx^(-alpha)/gamma(4-alpha);
#
# Independent variable for ODE integration
  t0=0;tf=0.1;nt=6;dt=(tf-t0)/(nt-1);
  tout=seq(from=t0,to=tf,by=dt);
#
# a_jk coefficients
  A=matrix(0,nrow=nx-2,ncol=nx-1);
  for(j in 1:(nx-2)){
    for(k in 0:j){
    if (k==0){
      A[j,k+1]=(j-1)^(3-alpha)-j^(2-alpha)*(j-3+alpha);
    } else if (1 <= k && k<=j-1){
      A[j,k+1]=(j-k+1)^(3-alpha)-2*(j-k)^(3-alpha)+(j-k-1)^(3-
          alpha);
    } else
      A[j,k+1]=1;
    }
  }
#
# Initial condition
  u0=rep(0,nx-2);
  for(j in 1:(nx-2)){
    u0[j]=f(xj[j+1]);}
  ncall=0;
#
# ODE integration
  out=lsode(y=u0,times=tout,func=pde1a,
      rtol=1e-6,atol=1e-6,maxord=5);
  nrow(out)
  ncol(out)
```

```
#
# Allocate array for u(x,t)
  u=matrix(0,nt,nx);
#
# u(x,t), x ne xl,xu
  for(i in 1:nt){
    for(j in 2:(nx-1)){
      u[i,j]=out[i,j];
    }
  }
#
# ncase=1 (with analytical solution)
  if(ncase==1){
#
# Reset boundary values
  for(i in 1:nt){
   u[i,1]=ua(xl,tout[i]);
  u[i,nx]=ua(xu,tout[i]);
  }
#
# Numerical, analytical solutions, max difference
  uap=matrix(0,nt,nx);
  for(i in 1:nt){
    for(j in 1:nx){
      uap[i,j]=ua((j-1)*dx,(i-1)*dt);
    }
  }
  max_err=max(abs(u-uap));
  }
#
# Tabular numerical, analytical solutions,
# difference
  cat(sprintf("\n       t      x     u(x,t)   ua(x,t)        diff"))
    ;
  for(i in 1:nt){
  iv=seq(from=1,to=nx,by=4);
  for(j in iv){
    cat(sprintf("\n %6.2f%6.2f%10.5f%10.5f%12.3e",
      tout[i],xj[j],u[i,j],uap[i,j],u[i,j]-uap[i,j]));
  }
```

```
  cat(sprintf("\n"));
  }
#
# Plot numerical, analytical solutions
  matplot(xj,t(u),type="l",lwd=2,col="black",lty=1,
    xlab="x",ylab="u(x,t)",main="");
  matpoints(xj,t(uap),pch="o",col="black");
#
# Display maximum error
  cat(sprintf("\n   maximum error = %6.2e \n",max_err));
#
# Plot error at t = tf
  err_1=abs(u[nt,]-ua(xj[1:nx],tf));
  plot(xj,err_1,type="l",xlab="x",
       ylab="Max Error at t = tf",
       main="",col="black")
#
# End ncase=1
  }
#
# ncase=2 (without analytical solution)
  if(ncase==2){
#
# Reset boundary values
  for(i in 1:nt){
   u[i,1]=0;
  u[i,nx]=0;
  }
#
# Tabular numerical solutions
# difference
  cat(sprintf("\n       t      x     u(x,t)"));
  for(i in 1:nt){
  iv=seq(from=1,to=nx,by=4);
  for(j in iv){
    cat(sprintf("\n %6.2f%6.2f%10.5f",
      tout[i],xj[j],u[i,j]));
  }
  cat(sprintf("\n"));
```

```
  }
#
# Plot numerical solution
  matplot(xj,t(u),type="l",lwd=2,col="black",lty=1,
    xlab="x",ylab="u(x,t)",main="");
#
# End ncase=2
  }
#
# Calls to ODE routine
  cat(sprintf("\n\n   ncall = %3d\n",ncall));
```

We can note the following details about Listing 8.1.

- Brief documentation comments are followed by deletion of previous files.

```
#
# Diffusion equation
#
#   ut=D*(d^alpha u/dx^alpha)
#
#   xl < x < xu, 0 < t < tf, xl=0, xu=1, D=1
#
#   u(x,t=0)=sin(pi*x)
#
#   u(x=xl,t)=u(x=xu,t)=0
#
#   ua(x,t)=exp(-pi^2*t)*sin(pi*x) (alpha=2)
#
# Delete previous workspaces
  rm(list=ls(all=TRUE))
```

- The ODE integrator library deSolve is accessed. The ODE/MOL routines is pde1a.

```
#
# Access functions for numerical solution
  library("deSolve");
  setwd("f:/fractional/sfpde/chap8/ex1/dirichlet");
  source("pde1a.R");
```

- Two cases are programmed. For `ncase=1` with `alpha=2`, the solution for the ISPDE is used to verify the numerical solution. For `ncase=2` with `alpha=1.5`, the effect of a variation in the fractional order α can be observed.

```
#
# Parameters
  ncase=1;
  if(ncase==1){alpha=2};
  if(ncase==2){alpha=1.5};
  D=1;
```

- A function for the analytical solution of eq. (8.1e) is programmed.

```
#
# Analytical solution
  ua=function(x,t) exp(-pi^2*t)*sin(pi*x);
```

- The IC of eq. (8.1b) follows from the analytical solution of eq. (8.1e) with $t = 0$.

```
#
# Initial condition function (IC)
  f=function(x) sin(pi*x);
```

- Homogeneous BCs (8.1c,d) are based on eqs. (3.4a,b).

```
#
# Boundary condition functions (BCs)
  g_0=function(t) 0;
  g_L=function(t) 0;
#
# Boundary condition coefficients
  c_1=function(t) 1;
  c_2=function(t) 0;
  c_3=function(t) 1;
  c_4=function(t) 0;
```

- A spatial grid of 41 points is defined for the interval $x = x_l = 0 \leq x \leq x = x_u = 1$, so that grid values are $x = 0, 0.025, ..., 1$.

```
#
# Spatial grid
```

```
    xl=0;xu=1;nx=41;dx=(xu-xl)/(nx-1);
    xj=seq(from=xl,to=xu,by=dx);
    cd=dx^(-alpha)/gamma(4-alpha);
```

- The interval in t, $t = t_0 = 0 \leq t \leq t = t_f = 0.1$, is defined with 6 points, so the output values are $t = 0, 0.02, ..., 0.1$.

```
#
# Independent variable for ODE integration
    t0=0;tf=0.1;nt=6;dt=(tf-t0)/(nt-1);
    tout=seq(from=t0,to=tf,by=dt);
```

- The coefficient matrix A is defined according to eq. (1.2g).

- IC (8.1b) follows from analytical solution (8.1e).

```
#
# Initial condition
    u0=rep(0,nx-2);
    for(j in 1:(nx-2)){
        u0[j]=f(xj[j+1]);}
    ncall=0;
```

- The 39 ODEs at the interior points in x are integrated with lsode.

```
#
# ODE integration
    out=lsode(y=u0,times=tout,func=pde1a,
        rtol=1e-6,atol=1e-6,maxord=5);
    nrow(out)
    ncol(out)
```

The ODE/MOL routine pde1a is considered next. The solution matrix out has dimensions $6 \times 39 + 1 = 40$ as confirmed in the subsequent output (the offset $+1$ provides for the value of t for each of the 6 derivative vectors).

- The numerical solution out returned by lsode is placed in matrix u (nx=41).

```
#
# Allocate array for u(x,t)
    u=matrix(0,nt,nx);
#
```

```
# u(x,t), x ne xl,xu
  for(i in 1:nt){
    for(j in 2:(nx-1)){
      u[i,j]=out[i,j];
    }
  }
```

- For ncase=1, the boundary values at $x = x_l, x_u$ are set by the analytical solution of eq. (8.1e).

```
#
# ncase=1 (with analytical solution)
  if(ncase==1){
#
# Reset boundary values
  for(i in 1:nt){
   u[i,1]=ua(xl,tout[i]);
   u[i,nx]=ua(xu,tout[i]);
   }
```

These are zero for $x_l = 0, x_u = 1$.

- The analytical solution of eq. (8.1e) is placed in array uap as a function of x and t, and the maximum absolute differece between the numerical and analytical solutions is placed in max_err.

```
#
# Numerical, analytical solutions, max difference
  uap=matrix(0,nt,nx);
  for(i in 1:nt){
    for(j in 1:nx){
      uap[i,j]=ua((j-1)*dx,(i-1)*dt);
    }
  max_err=max(abs(u-uap));
  }
```

Note that max and abs operate on matrices.

- The numerical and analytical solutions and the difference is displayed (every fourth value in x from by=4).

```
#
# Tabular numerical, analytical solutions,
# difference
  cat(sprintf("\n      t      x      u(x,t)    ua(x,t)        diff"));
  for(i in 1:nt){
  iv=seq(from=1,to=nx,by=4);
  for(j in iv){
    cat(sprintf("\n %6.2f%6.2f%10.5f%10.5f%12.3e",
      tout[i],xj[j],u[i,j],uap[i,j],u[i,j]-uap[i,j]));
  }
  cat(sprintf("\n"));
  }
```

• The numerical solution is plotted with `matplot` and the analytical solution is superimposed with `matpoints`.

```
#
# Plot numerical, analytical solutions
  matplot(xj,t(u),type="l",lwd=2,col="black",lty=1,
    xlab="x",ylab="u(x,t)",main="");
  matpoints(xj,t(uap),pch="o",col="black");
```

The transposes `t(u),t(uap)` are required so that the number of rows of `u,uap` equals the number of elements of `xj`. The solutions are plotted parametrically in t.

• The maximum difference of the numerical and analytical solutions is displayed. The error at $t = t_f$ is then plotted as a function of x.

```
#
# Display maximum error
  cat(sprintf("\n   maximum error = %6.2e \n",max_err));
#
# Plot error at t = tf
  err_1=abs(u[nt,]-ua(xj[1:nx],tf));
  plot(xj,err_1,type="l",xlab="x",
      ylab="Max Error at t = tf",
      main="",col="black")
#
# End ncase=1
  }
```

ncase=1 is terminated with the final }.

- For ncase=2 ($\alpha = 1.5$), and an analytical solution is not used.

```
#
# ncase=2 (without analytical solution)
  if(ncase==2){
#
# Reset boundary values
  for(i in 1:nt){
   u[i,1]=0;
  u[i,nx]=0;
  }
```

Homogeneous BCs (8.1c,d) are programmed directly.

- The numerical solution is displayed (every fourth value in x with by=4).

```
#
# Tabular numerical solutions
# difference
  cat(sprintf("\n        t     x     u(x,t)"));
  for(i in 1:nt){
  iv=seq(from=1,to=nx,by=4);
  for(j in iv){
    cat(sprintf("\n %6.2f%6.2f%10.5f",
      tout[i],xj[j],u[i,j]));
  }
  cat(sprintf("\n"));
  }
```

- The numerical solution is plotted with matplot and ncase=2 is concluded.

```
#
# Plot numerical solution
  matplot(xj,t(u),type="l",lwd=2,col="black",lty=1,
    xlab="x",ylab="u(x,t)",main="");
#
# End ncase=2
  }
```

- The number of calls to the ODE/MOL routine `pde1a` for the complete solution is displayed.

```
#
# Calls to ODE routine
  cat(sprintf("\n\n   ncall = %3d\n",ncall));
```

`pde1a` called by `lsode` is considered next.

8.2.2   ODE/MOL ROUTINE

Listing 8.2: ODE/MOL routine for eqs. (8.1).

```
  pde1a=function(t,u,parms){
#
# Function pde1a computes the derivative
# vector of the ODEs approximating the
# PDE
#
# Allocate the vector of the ODE
# derivatives
  nx=nx-2;
  ut=rep(0,nx);
#
# Approximation of uxx
  uxx=NULL;
#
# x=0
  u0=(2*dx*g_0(t)-c_2(t)*(4*u[1]-u[2]))/
     (2*dx*c_1(t)-3*c_2(t));
  uxx_0=2*u0-5*u[1]+4*u[2]-u[3];
  uxx[1]=u[2]-2*u[1]+u0;
#
# x=1
  un=(2*dx*g_L(t)+c_4(t)*(4*u[nx]-u[nx-1]))/
     (2*dx*c_3(t)+3*c_4(t));
  uxx[nx]=un-2*u[nx]+u[nx-1];
#
# Interior approximation of uxx
  for(k in 2:(nx-1)){
    uxx[k]=u[k+1]-2*u[k]+u[k-1];
```

```
    }
#
# PDE
#
# Step through ODEs
  for(j in 1:nx){
#
#   First term in series approximation of
#   fractional derivative
    ut[j]=A[j,1]*uxx_0;
#
#   Subsequent terms in series approximation
#   of fractional derivative
    for(k in 1:j){
      ut[j]=ut[j]+A[j,k+1]*uxx[k];
#
#   Next k (next term in series)
    }
    ut[j]=cd*D*ut[j];
#
# Next j (next ODE)
  }
#
# Increment calls to pde1a
  ncall <<- ncall+1;
#
# Return derivative vector of ODEs
  return(list(c(ut)));
  }
```

We can note the following details about Listing 8.2.

- The function is defined. Then a vector for the derivatives in t computed in pde1a is allocated. This vector applies to the MOL/ODEs at the $41 - 2 = 39$ interior points in x

```
    pde1a=function(t,u,parms){
#
# Function pde1a computes the derivative
# vector of the ODEs approximating the
# PDE
#
```

```
# Allocate the vector of the ODE
# derivatives
  nx=nx-2;
  ut=rep(0,nx);
```

- The derivative $\dfrac{\partial^2 u(x = \Delta x, t)}{\partial x^2}$ =uxx[1] is approximated with a finite difference (FD) (at the first interior point adjacent to $x = 0$). The FD approxcimation follows from eq. (1.2h).

```
#
# Approximation of uxx
  uxx=NULL;
#
# x=0
  u0=(2*dx*g_0(t)-c_2(t)*(4*u[1]-u[2]))/
     (2*dx*c_1(t)-3*c_2(t));
  uxx_0=2*u0-5*u[1]+4*u[2]-u[3];
  uxx[1]=u[2]-2*u[1]+u0;
```

The boundary value u0 at $x = 0$ follows from eq. (3.5c). The second derivative at $x = 0$, uxx_0, follows from eq. (1.2i) and is used in the subsequent approximation of the fractional derivative.

- un, uxx[nx] at $x = x_u$ are analogous to u0, uxx[1].

```
#
# x=1
  un=(2*dx*g_L(t)+c_4(t)*(4*u[nx]-u[nx-1]))/
     (2*dx*c_3(t)+3*c_4(t));
  uxx[nx]=un-2*u[nx]+u[nx-1];
```

- $\dfrac{\partial^2 u(x,t)}{\partial x^2}$ at $x = x_l + 2\Delta x, ..., x = x_u - 2\Delta x$ follows from eq. (1.2h).

```
#
# Interior approximation of uxx
  for(k in 2:(nx-1)){
    uxx[k]=u[k+1]-2*u[k]+u[k-1];
  }
```

- The fractional derivative in eq. (8.1a) is based on eq. (1.2j).

```
#
# PDE
#
# Step through ODEs
  for(j in 1:nx){
#
#    First term in series approximation of
#    fractional derivative
     ut[j]=A[j,1]*uxx_0;
#
#    Subsequent terms in series approximation
#    of fractional derivative
     for(k in 1:j){
       ut[j]=ut[j]+A[j,k+1]*uxx[k];
#
#    Next k (next term in series)
     }
```

- The MOL/ODEs approximating eq. (8.1a) are programmed.

```
     ut[j]=cd*D*ut[j];
#
# Next j (next ODE)
  }
```

The nx MOL/ODEs at the interior points follow from the completion of the for.

```
#
# Step through ODEs
  for(j in 1:nx){
```

- The counter for the calls to pde1a is incremented and returned to the main program of Listing 8.1 via <<-.

```
#
# Increment calls to pde1a
  ncall <<- ncall+1;
```

- The t derivative vector ut is returned as a list to lsode.

```
#
```

```
# Return derivative vector of ODEs
  return(list(c(ut)));
  }
```

c is the R vector operator. The final } concludes pde1a.

The numerical and graphical output from the routines of Listings 8.1, 8.2 is discussed next.

8.2.3   MODEL OUTPUT

Abbreviated numerical output is given in Table 8.1.

We can note the following details about the output in Table 8.1.

- The solution array out is $6 \times 39 + 1 = 40$ for the 39 ODEs at the interior points in x.

  [1] 6

  [1] 40

- The numerical and analytical solutions are identical at $t = 0$ since these ICs are set by the analytical solution at $t = 0$ (eq. (8.1e)). This is an important check since if the ICs are in error, the remainder of the solutions will be in error.

- BCs (8.1c,d) are confirmed.

- The overall accuracy of the numerical solution is indicated by the maximum error maximum error = 1.87e-04.

- The computational effort is modest, ncall = 266.

The agreement between the numerical and analytical solutions is confirmed in Figs. 8.1, 8.2.

This output is verification of the numerical algorithm and its computer implementation for the fractional derivative of eq. (8.1a).

For ncase=2, an analytical solution is not used (not readily available) for verification of the numerical solution. But the change in the solution from $\alpha = 2$ (ncase=1) to $\alpha = 1.5$ gives an indication of the effect of the order of the fractional derivative in eq. (8.1a).

Abbreviated numerical output is shown in Table 8.2.

The graphical ouput is in Fig. 8.3.

Figure 8.3 indicates that the change from $\alpha = 2$ to $\alpha = 1.5$ moves the system of eqs. (8.1) from totally parabolic to hyperbolic-parabolic, with an effective convective velocity that is negative so that the solution moves right to left.

Table 8.1: Abbreviated numerical output for eqs. (8.1), ncase=1 (*Continues.*)

[1] 6

[1] 40

t	x	u(x,t)	ua(x,t)	diff
0.00	0.00	0.00000	0.00000	0.000e+00
0.00	0.10	0.30902	0.30902	0.000e+00
0.00	0.20	0.58779	0.58779	0.000e+00
0.00	0.30	0.80902	0.80902	0.000e+00
0.00	0.40	0.95106	0.95106	0.000e+00
0.00	0.50	1.00000	1.00000	0.000e+00
0.00	0.60	0.95106	0.95106	0.000e+00
0.00	0.70	0.80902	0.80902	0.000e+00
0.00	0.80	0.58779	0.58779	0.000e+00
0.00	0.90	0.30902	0.30902	0.000e+00
0.00	1.00	0.00000	0.00000	0.000e+00
0.02	0.00	0.00000	0.00000	0.000e+00
0.02	0.10	0.25369	0.25366	2.553e-05
0.02	0.20	0.48254	0.48249	4.857e-05
0.02	0.30	0.66416	0.66410	6.685e-05
0.02	0.40	0.78077	0.78069	7.858e-05
0.02	0.50	0.82095	0.82087	8.263e-05
0.02	0.60	0.78077	0.78069	7.858e-05
0.02	0.70	0.66416	0.66410	6.685e-05
0.02	0.80	0.48254	0.48249	4.857e-05
0.02	0.90	0.25369	0.25366	2.553e-05
0.02	1.00	0.00000	0.00000	0.000e+00

.
.
.

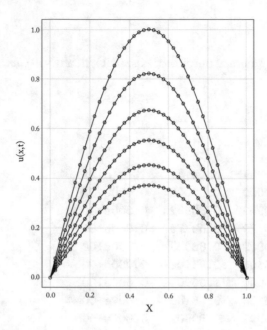

Figure 8.1: Numerical, analytical solutions of eqs. (8.1).

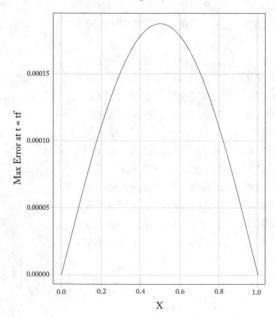

Figure 8.2: Error in the numerical solution of eqs. (8.1) at $t = t_f$.

Table 8.1: (*Continued.*) Abbreviated numerical output for eqs. (8.1), ncase=1

```
Output for t = 0.04, 0.06, 0.08 removed
          .                    .
          .                    .
          .                    .
 0.10  0.00   0.00000   0.00000   0.000e+00
 0.10  0.10   0.11523   0.11517   5.792e-05
 0.10  0.20   0.21918   0.21907   1.102e-04
 0.10  0.30   0.30168   0.30153   1.516e-04
 0.10  0.40   0.35464   0.35447   1.783e-04
 0.10  0.50   0.37290   0.37271   1.874e-04
 0.10  0.60   0.35464   0.35447   1.783e-04
 0.10  0.70   0.30168   0.30153   1.516e-04
 0.10  0.80   0.21918   0.21907   1.102e-04
 0.10  0.90   0.11523   0.11517   5.792e-05
 0.10  1.00   0.00000   0.00000   0.000e+00

maximum error = 1.87e-04

ncall = 266
```

To reiterate as a final conclusion, the numerical MOL algorithm for the fractional derivative of eqs. (8.1) is verified.

We now consider the preceding example for homogeneous Neumann BCs

$$\frac{\partial u(x = x_l, t)}{\partial x} = \frac{\partial u(x = x_u, t)}{\partial x} = 0 \qquad (8.2a,b)$$

The integer exact solution is

$$u_a(x, t) = e^{-\pi^2 t} \cos(\pi x) \qquad (8.2c)$$

which provides the IC

$$u_a(x, t = 0) = \cos(\pi x) \qquad (8.2d)$$

Equations (8.1a) and (8.2) constitute the test problem for Neumann BCs and are implemented in the following routines.

8.2.4   MAIN PROGRAM, NEUMANN BCS

The main program is next.

Table 8.2: Abbreviated numerical output for eq. (8.1a), ncase=2 (*Continues.*)

[1] 6

[1] 40

t	x	u(x,t)
0.00	0.00	0.99999
0.00	0.10	0.95106
0.00	0.20	0.80902
0.00	0.30	0.58779
0.00	0.40	0.30902
0.00	0.50	0.00000
0.00	0.60	-0.30902
0.00	0.70	-0.58779
0.00	0.80	-0.80902
0.00	0.90	-0.95106
0.00	1.00	-0.99999
0.02	0.00	0.94604
0.02	0.10	0.87727
0.02	0.20	0.71997
0.02	0.30	0.49590
0.02	0.40	0.22591
0.02	0.50	-0.06436
0.02	0.60	-0.34700
0.02	0.70	-0.59465
0.02	0.80	-0.78331
0.02	0.90	-0.89562
0.02	1.00	-0.92974

.
.
.

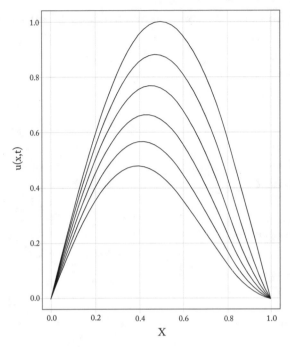

Figure 8.3: Numerical solutions of eqs. (8.1a), `ncase=2`.

Listing 8.3: Main program for eqs. (8.1a), (8.2), Neumann BCs.

```
#
# Fractional diffusion equation
#
#   ut=d*(d^alpha u/dx^alpha)
#
#   xl < x < xu, 0 < t < tf, xl=0, xu=1
#
#   u(x,t=0)=cos(pi*x)
#
#   ux(x=xl,t)=ux(x=xu,t)=0
#
#   ua(x,t)=exp(-pi^2*t)*cos(pi*x) (alpha=2)
#
# Delete previous workspaces
  rm(list=ls(all=TRUE))
#
```

Table 8.2: (*Continued.*) Abbreviated numerical output for eq. (8.1a), ncase=2

```
Output for t = 0.04,
0.06, 0.08 removed
          .
          .
          .

0.10   0.00     0.62715
0.10   0.10     0.54808
0.10   0.20     0.38947
0.10   0.30     0.19077
0.10   0.40    -0.02437
0.10   0.50    -0.23435
0.10   0.60    -0.41998
0.10   0.70    -0.56692
0.10   0.80    -0.66798
0.10   0.90    -0.72371
0.10   1.00    -0.74044

ncall = 273
```

```
# Access functions for numerical solution
  library("deSolve");
  setwd("f:/fractional/sfpde/chap8/ex1/neumann");
  source("pde1a.R");
#
# Parameters
  ncase=2;
  if(ncase==1){alpha=2};
  if(ncase==2){alpha=1.5};
  D=1;
#
# Analytical solution
  ua=function(x,t) exp(-pi^2*t)*cos(pi*x);
#
# Initial condition function (IC)
  f=function(x) cos(pi*x);
#
```

```
# Boundary condition functions (BCs)
  g_0=function(t) 0;
  g_L=function(t) 0;
#
# Boundary condition coefficients
  c_1=function(t) 0;
  c_2=function(t) 1;
  c_3=function(t) 0;
  c_4=function(t) 1;
#
# Spatial grid
  xl=0;xu=1;nx=41;dx=(xu-xl)/(nx-1);
  xj=seq(from=xl,to=xu,by=dx);
  cd=dx^(-alpha)/gamma(4-alpha);
#
# Independent variable for ODE integration
  t0=0;tf=0.1;nt=6;dt=(tf-t0)/(nt-1);
  tout=seq(from=t0,to=tf,by=dt);
#
# a_jk coefficients
  A= matrix(0,nrow=nx-2,ncol=nx-1);
  for(j in 1:(nx-2)){
    for(k in 0:j){
    if (k==0){
      A[j,k+1]=(j-1)^(3-alpha)-j^(2-alpha)*(j-3+alpha);
    } else if (1 <= k && k<=j-1){
      A[j,k+1]=(j-k+1)^(3-alpha)-2*(j-k)^(3-alpha)+(j-k-1)^(3-
          alpha);
    } else
      A[j,k+1]=1;
    }
  }
#
# Initial condition
  u0=rep(0,nx-2);
  for(j in 1:(nx-2)){
    u0[j]=f(xj[j+1]);}
  ncall=0;
#
```

```
# ODE integration
  out=lsode(y=u0,times=tout,func=pde1a,
      rtol=1e-6,atol=1e-6,maxord=5);
  nrow(out)
  ncol(out)
#
# Allocate array for u(x,t)
  u=matrix(0,nt,nx);
#
# u(x,t), x ne xl,xu
  for(i in 1:nt){
    for(j in 2:(nx-1)){
      u[i,j]=out[i,j];
    }
  }
#
# ncase=1 (with analytical solution)
  if(ncase==1){
#
# Reset boundary values
  for(i in 1:nt){
   u[i,1]=ua(xl,tout[i]);
  u[i,nx]=ua(xu,tout[i]);
  }
#
# Numerical, analytical solutions, max difference
  uap=matrix(0,nt,nx);
  for(i in 1:nt){
    for(j in 1:nx){
      uap[i,j]=ua((j-1)*dx,(i-1)*dt);
    }
  max_err=max(abs(u-uap));
  }
#
# Tabular numerical, analytical solutions,
# difference
  cat(sprintf("\n        t     x     u(x,t)    ua(x,t)        diff"))
    ;
  for(i in 1:nt){
```

```
  iv=seq(from=1,to=nx,by=4);
  for(j in iv){
    cat(sprintf("\n %6.2f%6.2f%10.5f%10.5f%12.3e",
      tout[i],xj[j],u[i,j],uap[i,j],u[i,j]-uap[i,j]));
  }
  cat(sprintf("\n"));
  }
#
# Plot numerical, analytical solutions
  matplot(xj,t(u),type="l",lwd=2,col="black",lty=1,
    xlab="x",ylab="u(x,t)",main="");
  matpoints(xj,t(uap),pch="o",col="black");
#
# Display maximum error
  cat(sprintf("\n   maximum error = %6.2e \n",max_err));
#
# Plot error at t = tf
  err_1=abs(u[nt,]-ua(xj[1:nx],tf));
  plot(xj,err_1,type="l",xlab="x",
      ylab="Max Error at t = tf",
      main="",col="black")
#
# End ncase=1
  }
#
# ncase=2 (without analytical solution)
  if(ncase==2){
#
# Reset boundary values
  for(i in 1:nt){
    u[i,1]=(2*dx*g_0(t)-c_2(t)*(4*u[i,2]-u[i,3]))/
          (2*dx*c_1(t)-3*c_2(t));
    u[i,nx]=(2*dx*g_L(t)+c_4(t)*(4*u[i,nx-1]-u[i,nx-2]))/
          (2*dx*c_3(t)+3*c_4(t));
  }
#
# Tabular numerical solutions
# difference
  cat(sprintf("\n     t     x     u(x,t)"));
```

```
  for(i in 1:nt){
  iv=seq(from=1,to=nx,by=4);
  for(j in iv){
    cat(sprintf("\n %6.2f%6.2f%10.5f",
      tout[i],xj[j],u[i,j]));
  }
  cat(sprintf("\n"));
  }
#
# Plot numerical solution
  matplot(xj,t(u),type="l",lwd=2,col="black",lty=1,
    xlab="x",ylab="u(x,t)",main="");
#
# End ncase=2
  }
#
# Calls to ODE routine
  cat(sprintf("\n\n   ncall = %3d\n",ncall));
```

This main program is similar to the main program of Listing 8.1 so only the differences are mentioned here.

- The analytical solution of eq. (8.2c) is programmed.

```
#
# Analytical solution
  ua=function(x,t) exp(-pi^2*t)*cos(pi*x);
```

- IC (8.2d) follows from the analytical solution with $t = 0$.

```
#
# Initial condition function (IC)
  f=function(x) cos(pi*x);
```

- The coefficients for Neumann BCs (8.2a,b) are defined.

```
#
# Boundary condition coefficients
  c_1=function(t) 0;
  c_2=function(t) 1;
  c_3=function(t) 0;
  c_4=function(t) 1;
```

8.2.5   ODE/MOL ROUTINE

The MOL/ODE routine `pde1a` is the same as in Listing 8.2 and therefore is not repeated here.

8.2.6   MODEL OUTPUT

Abbreviated numerical output is given in Table 8.3.

We can note the following details about the output in Table 8.3.

- The solution array out is $6 \times 39 + 1 = 40$ for the 39 ODEs at the interior points in x.

  ```
  [1] 6
  ```

  ```
  [1] 40
  ```

- The numerical and analytical solutions are identical at $t = 0$ since these ICs are set by the analytical solution at $t = 0$ (eq. (8.2d)).

- BCs (8.2a,b) are confirmed.

- The overall accuracy of the numerical solution is indicated by the maximum error `maximum error = 1.22e-04`.

- The computational effort is modest, `ncall = 267`.

The agreement between the numerical and analytical solutions is confirmed in Figs. 8.4, 8.5.

This output is again verification of the numerical algorithm and its computer implementation for the fractional derivative of eq. (8.1a).

For `ncase=2`, an analytical solution is not used for verification of the numerical solution. But the change in the solution from $\alpha = 2$ (`ncase=1`) to $\alpha = 1.5$ gives an indication of the effect of the order of the fractional derivative in eq. (8.1a),

Abbreviated numerical output is shown in Table 8.4.

The graphical ouput is in Fig. 8.6.

Figure 8.6 indicates that the change from $\alpha = 2$ to $\alpha = 1.5$ moves the system of eqs. (8.1a), (8.2) from totally parabolic to hyperbolic-parabolic, with an effective convective velocity that is negative so that the solution moves right to left.

To reiterate as a final conclusion, the numerical MOL algorithm for the fractional derivative of eqs. (8.1a), (8.2) is verified.

We now consider the preceding example for homogeneous Robin BCs

$$c_2(t)\frac{\partial u(x = x_l, t)}{\partial x} + c_1(t)u(x = x_l, t) = 0$$

$$c_4(t)\frac{\partial u(x = x_u, t)}{\partial x} + c_3(t)u(x = x_u, t) = 0 \qquad \text{(8.3a,b)}$$

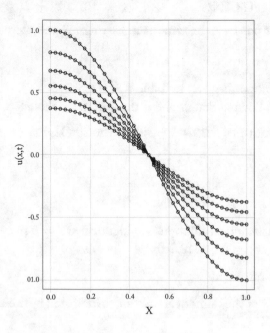

Figure 8.4: Numerical, analytical solutions of eqs. (8.1a), (8.2).

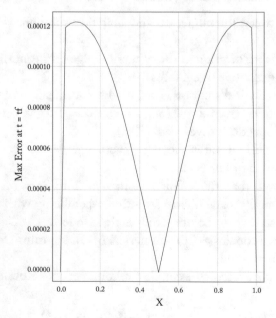

Figure 8.5: Error in the numerical solution of eqs. (8.1a), (8.2) at $t = t_f$.

Table 8.3: Abbreviated numerical output for eqs. (8.1a), (8.2) (*Continues.*)

[1] 6

[1] 40

t	x	u(x,t)	ua(x,t)	diff
0.00	0.00	1.00000	1.00000	0.000e+00
0.00	0.10	0.95106	0.95106	0.000e+00
0.00	0.20	0.80902	0.80902	0.000e+00
0.00	0.30	0.58779	0.58779	0.000e+00
0.00	0.40	0.30902	0.30902	0.000e+00
0.00	0.50	0.00000	0.00000	0.000e+00
0.00	0.60	−0.30902	−0.30902	0.000e+00
0.00	0.70	−0.58779	−0.58779	0.000e+00
0.00	0.80	−0.80902	−0.80902	0.000e+00
0.00	0.90	−0.95106	−0.95106	0.000e+00
0.00	1.00	−1.00000	−1.00000	0.000e+00
0.02	0.00	0.82087	0.82087	0.000e+00
0.02	0.10	0.78074	0.78069	5.120e−05
0.02	0.20	0.66415	0.66410	5.497e−05
0.02	0.30	0.48254	0.48249	4.433e−05
0.02	0.40	0.25369	0.25366	2.437e−05
0.02	0.50	−0.00000	0.00000	−5.154e−15
0.02	0.60	−0.25369	−0.25366	−2.437e−05
0.02	0.70	−0.48254	−0.48249	−4.433e−05
0.02	0.80	−0.66415	−0.66410	−5.497e−05
0.02	0.90	−0.78074	−0.78069	−5.120e−05
0.02	1.00	−0.82087	−0.82087	0.000e+00

.
.
.

Table 8.3: (*Continued.*) Abbreviated numerical output for eqs. (8.1a), (8.2)

```
   Output for t = 0.04, 0.06, 0.08 removed

       .                          .
       .                          .
       .                          .

   0.10  0.00    0.37271    0.37271    0.000e+00
   0.10  0.10    0.35459    0.35447    1.213e-04
   0.10  0.20    0.30164    0.30153    1.096e-04
   0.10  0.30    0.21915    0.21907    8.267e-05
   0.10  0.40    0.11522    0.11517    4.437e-05
   0.10  0.50   -0.00000    0.00000   -6.381e-14
   0.10  0.60   -0.11522   -0.11517   -4.437e-05
   0.10  0.70   -0.21915   -0.21907   -8.267e-05
   0.10  0.80   -0.30164   -0.30153   -1.096e-04
   0.10  0.90   -0.35459   -0.35447   -1.213e-04
   0.10  1.00   -0.37271   -0.37271    0.000e+00


   maximum error = 1.22e-04


   ncall = 267
```

with the IC

$$u(x, t = 0) = \sin(\pi x) \tag{8.3c}$$

Equations (8.1a) and (8.3) constitute the test problem for Robin BCs and are implemented in the following routines.

Table 8.4: Abbreviated numerical output for eqs. (8.1a), (8.2) ncase=2 (*Continues.*)

[1] 6

[1] 40

t	x	u(x,t)
0.00	0.00	0.00000
0.00	0.10	0.30902
0.00	0.20	0.58779
0.00	0.30	0.80902
0.00	0.40	0.95106
0.00	0.50	1.00000
0.00	0.60	0.95106
0.00	0.70	0.80902
0.00	0.80	0.58779
0.00	0.90	0.30902
0.00	1.00	0.00000
0.02	0.00	0.00000
0.02	0.10	0.29107
0.02	0.20	0.54474
0.02	0.30	0.73778
0.02	0.40	0.85249
0.02	0.50	0.87841
0.02	0.60	0.81355
0.02	0.70	0.66469
0.02	0.80	0.44713
0.02	0.90	0.19457
0.02	1.00	0.00000

.
.
.

Table 8.4: (*Continued.*) Abbreviated numerical output for eqs. (8.1a), (8.2) ncase=2

```
Output for t = 0.04,
0.06, 0.08 removed
          .
          .
          .

0.10   0.00    0.00000
0.10   0.10    0.19709
0.10   0.20    0.35225
0.10   0.30    0.44920
0.10   0.40    0.48020
0.10   0.50    0.44609
0.10   0.60    0.35882
0.10   0.70    0.24260
0.10   0.80    0.12846
0.10   0.90    0.04309
0.10   1.00    0.00000

ncall = 331
```

8.2.7   MAIN PROGRAM, ROBIN BCS

The main program is next.

Listing 8.4: Main program for eqs. (8.1a), (8.3).

```
#
# Fractional diffusion equation
#
#    ut=d*(d^alpha u/dx^alpha)
#
#    xl < x < xu, 0 < t < tf, xl=0, xu=1
#
#    u(x,t=0)=sin(pi*x)
#
#    ux(x=xl,t)-u(x=xl,t)=ux(x=xu,t)+u(x=xu,t)=0
#
# Delete previous workspaces
  rm(list=ls(all=TRUE))
#
```

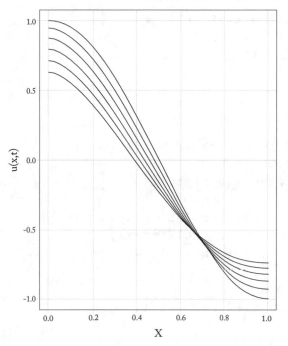

Figure 8.6: Numerical solutions of eqs. (8.1), (8.2), ncase=2.

```
# Access functions for numerical solution
  library("deSolve");
  setwd("f:/fractional/sfpde/chap8/ex1/robin");
  source("pde1a.R");
#
# Parameters
  ncase=1;
  if(ncase==1){alpha=2};
  if(ncase==2){alpha=1.5};
  D=1;
#
# Initial condition function (IC)
  f=function(x) sin(pi*x);
#
# Boundary condition functions (BCs)
  g_0=function(t) 0;
  g_L=function(t) 0;
#
```

```
# Boundary condition coefficients
  c_1=function(t)  -1;
  c_2=function(t)   1;
  c_3=function(t)   1;
  c_4=function(t)   1;
#
# Spatial grid
  xl=0;xu=1;nx=41;dx=(xu-xl)/(nx-1);
  xj=seq(from=xl,to=xu,by=dx);
  cd=dx^(-alpha)/gamma(4-alpha);
#
# Independent variable for ODE integration
  t0=0;tf=0.2;nt=6;dt=(tf-t0)/(nt-1);
  tout=seq(from=t0,to=tf,by=dt);
#
# a_jk coefficients
  A= matrix(0,nrow=nx-2,ncol=nx-1);
  for(j in 1:(nx-2)){
    for(k in 0:j){
    if (k==0){
      A[j,k+1]=(j-1)^(3-alpha)-j^(2-alpha)*(j-3+alpha);
    } else if (1 <= k && k<=j-1){
      A[j,k+1]=(j-k+1)^(3-alpha)-2*(j-k)^(3-alpha)+(j-k-1)^(3-
          alpha);
    } else
      A[j,k+1]=1;
    }
  }
#
# Initial condition
  u0=rep(0,nx-2);
  for(j in 1:(nx-2)){
    u0[j]=f(xj[j+1]);}
  ncall=0;
#
# ODE integration
  out=lsode(y=u0,times=tout,func=pde1a,
      rtol=1e-6,atol=1e-6,maxord=5);
  nrow(out)
```

```
  ncol(out)
#
# Allocate array for u(x,t)
  u=matrix(0,nt,nx);
#
# u(x,t), x ne xl,xu
  for(i in 1:nt){
    for(j in 2:(nx-1)){
      u[i,j]=out[i,j];
    }
  }
#
# Reset boundary values
  u[1,1]=0;u[1,nx]=0;
  for(i in 2:nt){
    u[i,1]=(2*dx*g_0(t)-c_2(t)*(4*u[i,2]-u[i,3]))/
           (2*dx*c_1(t)-3*c_2(t));
    u[i,nx]=(2*dx*g_L(t)+c_4(t)*(4*u[i,nx-1]-u[i,nx-2]))/
            (2*dx*c_3(t)+3*c_4(t));
  }
#
# Tabular numerical solutions
# difference
  cat(sprintf("\n       t      x     u(x,t)"));
  for(i in 1:nt){
  iv=seq(from=1,to=nx,by=4);
  for(j in iv){
    cat(sprintf("\n %6.2f%6.2f%10.5f",
      tout[i],xj[j],u[i,j]));
  }
  cat(sprintf("\n"));
  }
#
# Plot numerical solution
  matplot(xj,t(u),type="l",lwd=2,col="black",lty=1,
    xlab="x",ylab="u(x,t)",main="");
#
# Calls to ODE routine
  cat(sprintf("\n\n   ncall = %3d\n",ncall));
```

This main program is similar to the main program of Listing 8.1 so only the differences are mentioned here.

- IC is eq. (8.3c).

```
#
# Initial condition function (IC)
  f=function(x) sin(pi*x);
```

- The boundary condition coefficient for eqs. (8.3a,b) are specified.

```
#
# Boundary condition coefficients
  c_1=function(t)  -1;
  c_2=function(t)   1;
  c_3=function(t)   1;
  c_4=function(t)   1;
```

These coefficients were selected based on physical reasoning. Note in particular c_1(t)=-1, c_3(t)=1.

- The t interval has been extended from $t_f = 0.1$ to $t_f = 0.2$ to give a more complete solution.

- Since lsode does not return the dependent variable at the boundaries $x = x_l, x_u$ (only the solutions to ODEs are returned), the boundary values must be reset for subsequent display of the solution. In the two preceding examples (Dirichlet, Neumann BCs), this was done with the analytical solution. Since an analytical solution is not used in this (Robin) example, the boundary values are reset using the FD aproximations from pde1a. At $t = 0$, boundary values are used that are consistent with IC (8.3c).

```
#
# Reset boundary values
  u[1,1]=0;u[1,nx]=0;
  for(i in 2:nt){
    u[i,1]=(2*dx*g_0(t)-c_2(t)*(4*u[i,2]-u[i,3]))/
           (2*dx*c_1(t)-3*c_2(t));
    u[i,nx]=(2*dx*g_L(t)+c_4(t)*(4*u[i,nx-1]-u[i,nx-2]))/
            (2*dx*c_3(t)+3*c_4(t));
  }
```

The coding is the same for ncase=1,2 ($\alpha = 2, 1.5$) since an analytical solution is not used.

8.2.8   ODE/MOL ROUTINE

The MOL/ODE routine pde1a is the same as in Listing 8.2 and therefore is not repeated here.

8.2.9   MODEL OUTPUT

Abbreviated numerical output is given in Table 8.5.

We can note the following details about the output in Table 8.5.

- The solution array out is $6 \times 39 + 1 = 40$ for the 39 ODEs at the interior points in x.

  [1] 6

  [1] 40

- IC (8.3c) is confirmed.

- The solution is symmetric with respect to $t = 0.5$.

- The computational effort is modest, ncall = 865.

The graphical ouput is in Fig. 8.7.

Abbreviated numerical output for ncase=2 is shown in Table 8.6.

The graphical ouput is in Fig. 8.8.

Figure 8.8 indicates that the change from $\alpha = 2$ to $\alpha = 1.5$ moves the system of eqs. (8.1a), (8.3) from totally parabolic to hyperbolic-parabolic, with an effective convective velocity that is negative so that the solution moves right to left.

To reiterate, the solution from Listings 8.2, 8.4, ncase=1, is not verified with an analytical solution. The symmetry around $x = 0.5$ for $\alpha = 2$ is an essential check on the numerical solution.

Consideration is now given to a series of classical integer PDEs which are extended to fractical PDEs.

8.3   FRACTIONAL BURGERS EQUATION

The fractional Burgers equation is

$$\frac{\partial u}{\partial t} = v \frac{\partial^\alpha u}{\partial x^\alpha} - u \frac{\partial u}{\partial x} \tag{8.4a}$$

A MOL solution follows for the boundary conditions (BCs)

$$\frac{\partial u(x = 0, t)}{\partial x} = \frac{\partial u(x = 1, t)}{\partial x} = 0 \tag{8.4b,c}$$

and the initial condition (IC)

$$u(x, t = 0) = u_a(x, t = 0) \tag{8.4d}$$

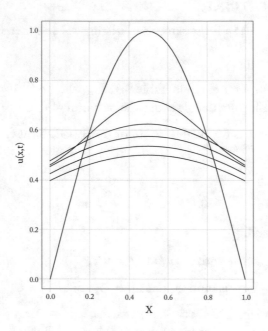

Figure 8.7: Numerical solution of eqs. (8.1a), (8.3), ncase=1.

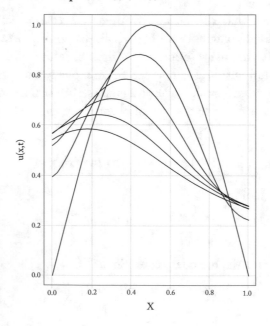

Figure 8.8: Numerical solution of eqs. (8.1a), (8.3), ncase=2.

Table 8.5: Abbreviated numerical output for eqs. (8.1a), (8.3), ncase=1 (*Continues.*)

```
[1]  6

[1]  40

       t      x      u(x,t)
     0.00   0.00    0.00000
     0.00   0.10    0.30902
     0.00   0.20    0.58779
     0.00   0.30    0.80902
     0.00   0.40    0.95106
     0.00   0.50    1.00000
     0.00   0.60    0.95106
     0.00   0.70    0.80902
     0.00   0.80    0.58779
     0.00   0.90    0.30902
     0.00   1.00    0.00000

     0.04   0.00    0.46061
     0.04   0.10    0.51549
     0.04   0.20    0.58353
     0.04   0.30    0.65135
     0.04   0.40    0.70162
     0.04   0.50    0.72019
     0.04   0.60    0.70162
     0.04   0.70    0.65135
     0.04   0.80    0.58353
     0.04   0.90    0.51549
     0.04   1.00    0.46061
              .
              .
              .
```

Table 8.5: (*Continued.*) Abbreviated numerical output for eqs. (8.1a), (8.3), ncase=1

```
Output for t = 0.08,
0.12, 0.16 removed
          .
          .
          .
0.20   0.00    0.39757
0.20   0.10    0.43383
0.20   0.20    0.46271
0.20   0.30    0.48372
0.20   0.40    0.49648
0.20   0.50    0.50076
0.20   0.60    0.49648
0.20   0.70    0.48372
0.20   0.80    0.46271
0.20   0.90    0.43383
0.20   1.00    0.39757


ncall = 865
```

The analytical solution for $\alpha = 2$ is [2]

$$u_a(x,t) = \frac{0.1e^{-A} + 0.5e^{-B} + e^{-C}}{e^{-A} + e^{-B} + e^{-C}} \qquad (8.4e)$$

where

$$A = \frac{0.05}{v}(x - 0.5 + 4.95t)$$

$$B = \frac{0.25}{v}(x - 0.5 + 0.75t)$$

$$C = \frac{0.5}{v}(x - 0.375)$$

MOL routines for eqs. (8.4) follow.

8.3.1   MAIN PROGRAM, DIRCHLET BCS

A main program for eqs. (8.4) is next.

Table 8.6: Abbreviated numerical output for eqs. (8.1), (8.3), ncase=2 (*Continues.*)

[1] 6

[1] 40

t	x	u(x,t)
0.00	0.00	0.00000
0.00	0.10	0.30902
0.00	0.20	0.58779
0.00	0.30	0.80902
0.00	0.40	0.95106
0.00	0.50	1.00000
0.00	0.60	0.95106
0.00	0.70	0.80902
0.00	0.80	0.58779
0.00	0.90	0.30902
0.00	1.00	0.00000
0.04	0.00	0.39492
0.04	0.10	0.51420
0.04	0.20	0.67574
0.04	0.30	0.80672
0.04	0.40	0.87526
0.04	0.50	0.86692
0.04	0.60	0.77911
0.04	0.70	0.62140
0.04	0.80	0.43049
0.04	0.90	0.28343
0.04	1.00	0.22387

.

.

.

Table 8.6: (*Continued.*) Abbreviated numerical output for eqs. (8.1), (8.3), ncase=2

```
Output for t = 0.08,
0.12, 0.16 removed
             .
             .
             .
0.20   0.00    0.54197
0.20   0.10    0.57695
0.20   0.20    0.58474
0.20   0.30    0.56812
0.20   0.40    0.53141
0.20   0.50    0.48221
0.20   0.60    0.42906
0.20   0.70    0.37880
0.20   0.80    0.33524
0.20   0.90    0.29923
0.20   1.00    0.26963


ncall = 432
```

Listing 8.5: Main program for eqs. (8.4).

```
#
# Fractional Burgers
#
#    ut=vis*(d^alpha u/dx^alpha)-u*ux
#
#    xl < x < xu, 0 < t < tf, xl=0, xu=1
#
#    u(x,t=0)=ua(x,t=0)
#
#    c1(t)*ux(x=xl,t)+c2(t)*u(x=xl,t)=0
#
#    c3(t)*ux(x=xu,t)+c4(t)*u(x=xu,t)=0
#
#    ua(x,t) (analytical solution a
#             multiline function)
#
```

```
# Delete previous workspaces
  rm(list=ls(all=TRUE))
#
# Access functions for numerical solution
  library("deSolve");
  setwd("f:/fractional/sfpde/chap8/ex2");
  source("pde1a.R");source("ua.R")
#
# Parameters
  ncase=1;
  if(ncase==1){alpha=2;}
  if(ncase==2){alpha=1.9;}
  vis=0.003;
#
# Boundary condition functions
  g_0=function(t) 0;
  g_L=function(t) 0;
#
# Boundary condition coefficients
  c_1=function(t) 0;
  c_2=function(t) 1;
  c_3=function(t) 0;
  c_4=function(t) 1;
#
# Spatial grid
  xl=0;xu=1;nx=101;dx=(xu-xl)/(nx-1);
  xj=seq(from=xl,to=xu,by=dx);
  cd=dx^(-alpha)/gamma(4-alpha);
  r12dx=1/(12*dx);
#
# Independent variable for ODE integration
  t0=0;tf=1;nt=11;dt=(tf-t0)/(nt-1);
  tout=seq(from=t0,to=tf,by=dt);
#
# a_jk coefficients
  A=matrix(0,nrow=nx-2,ncol=nx-1);
  for(j in 1:(nx-2)){
    for(k in 0:j){
    if (k==0){
```

```
      A[j,k+1]=(j-1)^(3-alpha)-j^(2-alpha)*(j-3+alpha);
    } else if (1 <= k && k<=j-1){
      A[j,k+1]=(j-k+1)^(3-alpha)-2*(j-k)^(3-alpha)+(j-k-1)^(3-
          alpha);
    } else
      A[j,k+1]=1;
    }
  }
#
# Initial condition
  u0=rep(0,nx-2);
  for(j in 1:(nx-2)){
    u0[j]=ua(xj[j+1],0);}
  ncall=0;
#
# ODE integration
  out=lsodes(func=pde1a,y=u0,times=tout,
             sparsetype ="sparseint")
  nrow(out)
  ncol(out)
#
# Allocate array for u(x,t)
  u=matrix(0,nt,nx);
#
# u(x,t), x ne xl,xu
  for(i in 1:nt){
    for(j in 2:(nx-1)){
      u[i,j]=out[i,j];
    }
  }
#
# Reset boundary values
  for(i in 1:nt){
   u[i,1]=ua(xl,tout[i]);
  u[i,nx]=ua(xu,tout[i]);
  }
#
# ncase=1
  if(ncase==1){
```

```
#
# Numerical, analytical solutions, difference
  uap=matrix(0,nt,nx);
  for(i in 1:nt){
    for(j in 1:nx){
      uap[i,j]=ua((j-1)*dx,(i-1)*dt);
    }
  max_err=max(abs(u-uap));
  }
#
# Tabular numerical, analytical solutions,
# difference
  cat(sprintf("\n       t      x     u(x,t)   ua(x,t)        diff"))
    ;
  for(i in 1:nt){
  iv=seq(from=1,to=nx,by=10);
  for(j in iv){
    cat(sprintf("\n %6.2f%6.2f%10.5f%10.5f%12.3e",
      tout[i],xj[j],u[i,j],uap[i,j],u[i,j]-uap[i,j]));
  }
  cat(sprintf("\n"));
  }
#
# Plot numerical, analytical solutions
  matplot(xj,t(u),type="l",lwd=2,col="black",lty=1,
    xlab="x",ylab="u(x,t)",main="");
  matpoints(xj,t(uap),pch="o",col="black");
#
# Display maximum error
  cat(sprintf("\n   maximum error = %6.2e \n",max_err));
#
# Plot error at t = tf
  err_1=abs(u[nt,]-ua(xj[1:nx],tf));
  plot(xj,err_1,type="l",xlab="x",
      ylab="Max Error at t = tf",
      main="",col="black")
  }
#
# ncase=2
```

```
  if(ncase==2){
#
# Tabular numerical solution
  cat(sprintf("\n        t      x      u(x,t)"));
  for(i in 1:nt){
  iv=seq(from=1,to=nx,by=10);
  for(j in iv){
    cat(sprintf("\n %6.2f%6.2f%10.5f",
      tout[i],xj[j],u[i,j]));
  }
  cat(sprintf("\n"));
  }
#
# Plot numerical solution
  matplot(xj,t(u),type="l",lwd=2,col="black",lty=1,
    xlab="x",ylab="u(x,t)",main="");
  }
#
# Calls to ODE routine
  cat(sprintf("\n\n  ncall = %3d\n",ncall));
```

We can note the following details about Listing 8.5.

- After brief documentation comments, previous workspaces are deleted.

```
  #
  # Fractional Burgers
  #
  #    ut=vis*(d^alpha u/dx^alpha)-u*ux
  #
  #    xl < x < xu, 0 < t < tf, xl=0, xu=1
  #
  #    u(x,t=0)=ua(x,t=0)
  #
  #    c1(t)*ux(x=xl,t)+c2(t)*u(x=xl,t)=0
  #
  #    c3(t)*ux(x=xu,t)+c4(t)*u(x=xu,t)=0
  #
  #    ua(x,t) (analytical solution a
  #              multiline function)
  #
```

```
# Delete previous workspaces
  rm(list=ls(all=TRUE))
```

- The ODE integrator library deSolve is accessed. The ODE/MOL routine is pde1a.

```
#
# Access functions for numerical solution
  library("deSolve");
  setwd("f:/fractional/sfpde/chap8/ex2");
  source("pde1a.R");source("ua.R")
```

The analytical solution of eq. (8.4e) is in file ua.

- Two cases are defined (and explained subsequently).

```
#
# Parameters
  ncase=1;
  if(ncase==1){alpha=2;}
  if(ncase==2){alpha=1.9;}
  vis=0.003;
```

The coefficient ν=vis in eq. (8.4a) is defined numerically.

- Boundary condition functions and coefficients are defined for Neumann BCs (8.4b,c).

```
#
# Boundary condition functions
  g_0=function(t) 0;
  g_L=function(t) 0;
#
# Boundary condition coefficients
  c_1=function(t) 0;
  c_2=function(t) 1;
  c_3=function(t) 0;
  c_4=function(t) 1;
```

- A spatial grid of 101 points is defined for the interval $x_l = 0 \le x \le x_u = 1$, so that grid values are $x = 0, 0.01, ..., 1$.

```
#
# Spatial grid
```

```
    xl=0;xu=1;nx=101;dx=(xu-xl)/(nx-1);
    xj=seq(from=xl,to=xu,by=dx);
    cd=dx^(-alpha)/gamma(4-alpha);
    r12dx=1/(12*dx);
```

- The interval in t, $t = t_0 = 0 \leq t \leq t = t_f = 1$, is defined with 11 points, so the output values are $t = 0, 0.1, ..., 1$.

```
#
# Independent variable for ODE integration
    t0=0;tf=1;nt=11;dt=(tf-t0)/(nt-1);
    tout=seq(from=t0,to=tf,by=dt);
```

- The A matrix of eqs. (1.2g) for the MOL analysis is defined.

```
#
# a_jk coefficients
    A=matrix(0,nrow=nx-2,ncol=nx-1);
    for(j in 1:(nx-2)){
      for(k in 0:j){
      if (k==0){
        A[j,k+1]=(j-1)^(3-alpha)-j^(2-alpha)*(j-3+alpha);
      } else if (1 <= k && k<=j-1){
        A[j,k+1]=(j-k+1)^(3-alpha)-2*(j-k)^(3-alpha)+(j-k-1)^(3-alpha);
      } else
        A[j,k+1]=1;
      }
    }
```

- IC (8.4d) is implemented via the analytical solution of eq. (8.4e) with $t = 0$.

```
#
# Initial condition
    u0=rep(0,nx-2);
    for(j in 1:(nx-2)){
      u0[j]=ua(xj[j+1],0);}
    ncall=0;
```

- The 99 ODEs at the interior points in x are integrated with lsodes.

```
#
```

```
# ODE integration
  out=lsodes(func=pde1a,y=u0,times=tout,
              sparsetype ="sparseint")
  nrow(out)
  ncol(out)
```

The ODE/MOL routine pde1a is considered next. The solution matrix out has dimensions $11 \times 99 + 1 = 100$ as confirmed in the subsequent output (the offset $+1$ provides for the value of t for each of the 11 derivative vectors).

lsodes was used (rather than, for example, lsode) since the ODE system has become relatively large (99 interior ODEs). The sparse matrix facility of lsodes is increasingly efficient with increasing size of the MOL/ODE system.

-
```
#
# Allocate array for u(x,t)
  u=matrix(0,nt,nx);
#
# u(x,t), x ne xl,xu
  for(i in 1:nt){
    for(j in 2:(nx-1)){
      u[i,j]=out[i,j];
    }
  }
```

- The solution in out from lsodes is placed in array u.

```
#
# Allocate array for u(x,t)
  u=matrix(0,nt,nx);
#
# u(x,t), x ne xl,xu
  for(i in 1:nt){
    for(j in 2:(nx-1)){
      u[i,j]=out[i,j];
    }
  }
```

- The boundary values of the solution are set by the analytical solution of eqs. (8.4e).

```
#
# Reset boundary values
```

```
for(i in 1:nt){
  u[i,1]=ua(xl,tout[i]);
u[i,nx]=ua(xu,tout[i]);
  }
```

The analytical solution of eq. (8.4e) gives boundary values corresponding to the homogenous Neumann BCs of eqs. (8.4b,c).

- For ncase=1 ($\alpha = 2$), the analytical solution of eq. (8.4e) is placed in array uap and the maximum difference between the numerical and analytical solutions is placed in max_err.

```
#
# ncase=1
  if(ncase==1){
#
# Numerical, analytical solutions, difference
  uap=matrix(0,nt,nx);
  for(i in 1:nt){
    for(j in 1:nx){
      uap[i,j]=ua((j-1)*dx,(i-1)*dt);
    }
  max_err=max(abs(u-uap));
  }
```

R functions max, abs operate on matrices.

- The numerical and analytical solution and the difference are displayed.

```
#
# Tabular numerical, analytical solutions,
# difference
  cat(sprintf("\n     t    x    u(x,t)   ua(x,t)      diff"));
  for(i in 1:nt){
  iv=seq(from=1,to=nx,by=10);
  for(j in iv){
    cat(sprintf("\n %6.2f%6.2f%10.5f%10.5f%12.3e",
      tout[i],xj[j],u[i,j],uap[i,j],u[i,j]-uap[i,j]));
  }
  cat(sprintf("\n"));
  }
```

Every tenth value of x appears from by=10.

- The numerical solution is plotted with `matplot` and the analytical solution is superimposed with `matpoints`. The transposes `t(u),t(uap)` are required so the dimensions of `xj,u,uap` are consistent (conformable). The solutions are plotted parametrically in t.

```
#
# Plot numerical, analytical solutions
  matplot(xj,t(u),type="l",lwd=2,col="black",lty=1,
    xlab="x",ylab="u(x,t)",main="");
  matpoints(xj,t(uap),pch="o",col="black");
```

- The maximum error is displayed and the error is plotted again x at $t = t_f$.

```
#
# Display maximum error
  cat(sprintf("\n   maximum error = %6.2e \n",max_err));
#
# Plot error at t = tf
  err_1=abs(u[nt,]-ua(xj[1:nx],tf));
  plot(xj,err_1,type="l",xlab="x",
       ylab="Max Error at t = tf",
       main="",col="black")
  }
```

The final } concludes ncase=1.

- For `ncase=2` ($\alpha = 2$), analytical solution (8.4e) does not apply so it is not included in the output.

```
#
# ncase=2
  if(ncase==2){
#
# Tabular numerical solution
  cat(sprintf("\n      t      x      u(x,t)"));
  for(i in 1:nt){
  iv=seq(from=1,to=nx,by=10);
  for(j in iv){
    cat(sprintf("\n %6.2f%6.2f%10.5f",
      tout[i],xj[j],u[i,j]));
  }
  cat(sprintf("\n"));
  }
```

- The numerical solution is plotted against x.

```
#
# Plot numerical solution
  matplot(xj,t(u),type="l",lwd=2,col="black",lty=1,
    xlab="x",ylab="u(x,t)",main="");
  }
```

The final } concludes ncase=2.

- ```
  #
  # Calls to ODE routine
    cat(sprintf("\n\n  ncall = %3d\n",ncall));
  ```

- The number of calls to pde1a is displayed at the end of each solution.

```
#
# Calls to ODE routine
  cat(sprintf("\n\n  ncall = %3d\n",ncall));
```

8.3.2   ODE/MOL ROUTINE

pde1a called by lsodes follows.

Listing 8.6: ODE/MOL routine for eqs. (8.4).

```
  pde1a=function(t,u,parms){
#
# Function pde1a computes the derivative
# vector of the ODEs approximating the
# PDE
#
# Allocate the vector of the interior ODEs
  nx=nx-2;
  ut=rep(0,nx);
#
# Boundary approximations of uxx
  uxx=NULL;
#
# x=0
  u0=(2*dx*g_0(t)-c_2(t)*(4*u[1]-u[2]))/
      (2*dx*c_1(t)-3*c_2(t));
  uxx_0=2*u0-5*u[1]+4*u[2]-u[3];
```
```

```
  uxx[1]=u[2]-2*u[1]+u0;
#
# x=1
  un=(2*dx*g_L(t)+c_4(t)*(4*u[nx]-u[nx-1]))/
     (2*dx*c_3(t)+3*c_4(t));
  uxx[nx]=un-2*u[nx]+u[nx-1];
#
# Interior approximation of uxx
  for(k in 2:(nx-1)){
    uxx[k]=u[k+1]-2*u[k]+u[k-1];
  }
#
# PDE
#
# Approximation of ux
  ux=NULL;
  ux[1]=r12dx*(-25*u[   1]+48*u[   2]-36*u[   3]+
                 16*u[   4] -3*u[   5]);
  ux[2]=r12dx*(  -3*u[   1]-10*u[   2]+18*u[   3]-
                  6*u[   4]   +u[   5]);
  ux[3]=r12dx*(     u[   1] -8*u[   2]+
                  8*u[   4]   -u[   5]);
  for(i in 4:(nx-1)){
    ux[i]=r12dx*(  -u[i-3] +6*u[i-2]-18*u[i-1]+
                 10*u[i  ] +3*u[i+1]);}
  ux[nx]=r12dx*(  3*u[nx-4]-16*u[nx-3]+36*u[nx-2]-
                 48*u[nx-1]+25*u[nx  ]);
#
# Step through ODEs
  for(j in 1:nx){
#
#   First term in series approximation of
#   fractional derivative
    ut[j]=A[j,1]*uxx_0;
#
#   Subsequent terms in series approximation
#   of fractional derivative
    for(k in 1:j){
      ut[j]=ut[j]+A[j,k+1]*uxx[k];
```

```
#
#    Next k (next term in series)
    }
    ut[j]=cd*vis*ut[j]-u[j]*ux[j];
#
# Next j (next ODE)
  }
#
# Increment calls to pde1a
  ncall <<- ncall+1;
#
# Return derivative vector of ODEs
  return(list(c(ut)));
  }
```

Listing 8.6 is similar to Listing 8.2 and only the differences are discussed.

- Since eq. (8.4a), has a nonlinear convective (hyperbolic) term $-u\dfrac{\partial u}{\partial x}$, an approximation for the first derivative is required. Also, eq. (8.4a) exhibits steep, moving fronts so that a two point upwind (2PU) FD is not sufficiently accurate. Therefore, a five point biased upwind (5PBU) FD ([3], pp 134-141) approximation is used.

```
#
# Approximation of ux
  ux=NULL;
  ux[1]=r12dx*(-25*u[  1]+48*u[  2]-36*u[  3]+
                16*u[  4] -3*u[  5]);
  ux[2]=r12dx*( -3*u[  1]-10*u[  2]+18*u[  3]-
                 6*u[  4]  +u[  5]);
  ux[3]=r12dx*(    u[  1] -8*u[  2]+
                 8*u[  4]   -u[  5]);
  for(i in 4:(nx-1)){
    ux[i]=r12dx*( -u[i-3] +6*u[i-2]-18*u[i-1]+
              10*u[i  ] +3*u[i+1]);}
  ux[nx]=r12dx*( 3*u[nx-4]-16*u[nx-3]+36*u[nx-2]-
              48*u[nx-1]+25*u[nx  ]);
```

This code requires some additional explanation.

– The programming is for the 99 interior points. At the left point, the first derivative, ux[1], is based on points 1,2,3,4,5, that is, only interior points or totally biased in the downwind direction to the right of point 1.

```
ux[1]=r12dx*(-25*u[  1]+48*u[  2]-36*u[  3]+
              16*u[  4] -3*u[  5]);
```

r12dx=$\dfrac{2}{4!\Delta x}$ is defined in the main program of Listing 8.5.

At the second interior point, the first derivative, ux[2], is again based on points 1,2,3,4,5.

```
ux[2]=r12dx*(  -3*u[  1]-10*u[  2]+18*u[  3]-
               6*u[  4]   +u[  5]);
```

– At the third point, the first derivative, ux[3, is based on points 1,2,4,5.

```
ux[3]=r12dx*(     u[  1] -8*u[  2]+
               8*u[  4]   -u[  5]);
```

Note that this FD is centered.

– At points i = 4,5,...,nx-1 = 4,5,...,98, the first derivative is based on one point to the right and three points to the left of point i. Therefore, the FD is biased in the upwind direction which for $v > 0$ is to the left of the point i, that is, three points to the left and one point to the right. In this case, the velocity is the dependent variable u[i] which is positive (in the product $-u\dfrac{\partial u}{\partial x}$).

```
for(i in 4:(nx-1)){
  ux[i]=r12dx*( -u[i-3] +6*u[i-2]-18*u[i-1]+
               10*u[i  ] +3*u[i+1]);}
```

Similar FDs follow for $v < 0$ which are biased to the right.

– Finally, at the point nx, the first derivative, ux[nx], is based on points nx,nx-1,nx-2,nx-3,nx-4, that is, it is totally biased in the upwind directon (for $v > 0$).

```
ux[nx]=r12dx*( 3*u[nx-4]-16*u[nx-3]+36*u[nx-2]-
              48*u[nx-1]+25*u[nx  ]);
```

– In this way, all of the interior ODE points, i = 1,2,...,nx = i=1,2,...,99, are accommodated.

• Equation (8.4a) is programmed as

```
ut[j]=cd*vis*ut[j]-u[j]*ux[j];
```

The ease of including the nonlinear convection term, $-u\dfrac{\partial u}{\partial x}$, is noteworthy.

The programming of the analytical solution of eq. (8.4e) is straightforward and does not require further discussion.

```
  ua=function(x,t){
#
# Funtion ua computes the exact solution of the
# Burgers equation for comparison with the numerical
# solution.
#
# Analytical solution
  a=(0.05/vis)*(x-0.5+4.95*t);
  b=(0.25/vis)*(x-0.5+0.75*t);
  c=( 0.5/vis)*(x-0.375);
  ea=exp(-a);
  eb=exp(-b);
  ec=exp(-c);
  ua=(0.1*ea+0.5*eb+ec)/(ea+eb+ec);
  return(c(ua));
}
```

This concludes the programming of eqs. (8.4). The output is considered subsequently.

8.3.3   MODEL OUTPUT

Abbreviated numerical output for `ncase=1` is shown in Table 8.7.

We can note the following details about the output in Table 8.7.

- The solution array out is $11 \times 99 + 1 = 100$ for the 99 ODEs at the interior points in x.

  [1] 11

  [1] 100

- The numerical and analytical solutions are the same at $t = 0$ as expected since both are set according to IC (8.4d).

- The maximum error, `maximum error = 4.95e-02`, generally occurs where the solution changes most rapidly, e.g.,

```
    1.00  0.80   1.00000   1.00000  -1.118e-08
    1.00  0.90   0.82751   0.85695  -2.944e-02
    1.00  1.00   0.10000   0.10000   0.000e+00
```

Table 8.7: Abbreviated numerical output for eqs. (8.4), ncase=1 (*Continues.*)

[1] 11

[1] 100

t	x	u(x,t)	ua(x,t)	diff
0.00	0.00	1.00000	1.00000	0.000e+00
0.00	0.10	1.00000	1.00000	0.000e+00
0.00	0.20	0.99237	0.99237	0.000e+00
0.00	0.30	0.50763	0.50763	0.000e+00
0.00	0.40	0.49949	0.49949	0.000e+00
0.00	0.50	0.30000	0.30000	0.000e+00
0.00	0.60	0.10051	0.10051	0.000e+00
0.00	0.70	0.10000	0.10000	0.000e+00
0.00	0.80	0.10000	0.10000	0.000e+00
0.00	0.90	0.10000	0.10000	0.000e+00
0.00	1.00	0.10000	0.10000	0.000e+00
0.10	0.00	1.00000	1.00000	0.000e+00
0.10	0.10	1.00000	1.00000	-2.845e-09
0.10	0.20	0.99998	0.99999	-1.867e-07
0.10	0.30	0.94420	0.94464	-4.385e-04
0.10	0.40	0.50109	0.50089	1.924e-04
0.10	0.50	0.45243	0.45232	1.063e-04
0.10	0.60	0.10385	0.10373	1.226e-04
0.10	0.70	0.10001	0.10000	3.066e-07
0.10	0.80	0.10000	0.10000	3.916e-10
0.10	0.90	0.10000	0.10000	4.983e-13
0.10	1.00	0.10000	0.10000	0.000e+00

·
·
·

Table 8.7: (*Continued.*) Abbreviated numerical output for eqs. (8.4), ncase=1

```
    Output for t = 0.2, ..., 0.9 removed
              .                    .
              .                    .
              .                    .

  1.00  0.00   1.00000   1.00000   0.000e+00
  1.00  0.10   1.00000   1.00000  -8.840e-09
  1.00  0.20   1.00000   1.00000  -8.840e-09
  1.00  0.30   1.00000   1.00000  -8.840e-09
  1.00  0.40   1.00000   1.00000  -8.840e-09
  1.00  0.50   1.00000   1.00000  -8.840e-09
  1.00  0.60   1.00000   1.00000  -8.840e-09
  1.00  0.70   1.00000   1.00000  -8.840e-09
  1.00  0.80   1.00000   1.00000  -1.118e-08
  1.00  0.90   0.82751   0.85695  -2.944e-02
  1.00  1.00   0.10000   0.10000   0.000e+00


maximum error = 4.95e-02

ncall = 539
```

However, as observed in Figs. 8.9, 8.10, this error does not invalidate the numerical solution.

- The computational effort is modest.

```
    ncall = 539
```

The graphical output is in Figs. 8.9, 8.10.

Figure 8.9 indicates the solution is acceptable, even for the steep moving fronts. Also, these fronts sharpen as t increases, indicating the numerical solution could eventually fail. This point is illustrated with ncase=2.

Abbreviated numerical output for ncase=2 is shown in Table 8.8.

Equations (8.4) moves from hyperbolic-parabolic with ncase=1, $\alpha = 2$, to a greater hyperbolic component from the fractional derivative with ncase=2, $\alpha = 1.9$. As expected, the moving front sharpens as reflected in Table 8.8 and Fig. 8.11. If α is reduced further, the numerical MOL algorithm fails. Thus, ncase=1,2 gives a clear indication of front sharpening for the Burgers eq. (8.4a) as it becomes more hyperbolic (convective).

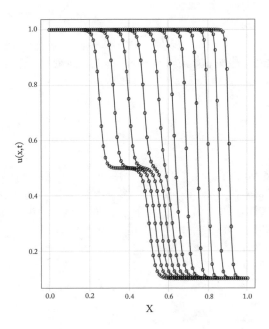

Figure 8.9: Numerical and analytical solutions of eqs. (8.4), `ncase=1`.

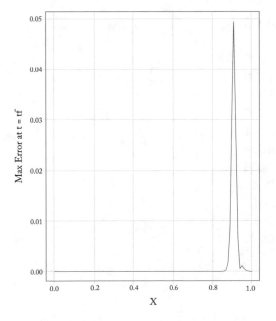

Figure 8.10: Error in numerical solution at $t = t_f = 1$, `ncase=1`.

Table 8.8: Abbreviated numerical output for eqs. (8.4), ncase=2 (*Continues.*)

[1] 11

[1] 100

t	x	u(x,t)
0.00	0.00	1.00000
0.00	0.10	1.00000
0.00	0.20	0.99237
0.00	0.30	0.50763
0.00	0.40	0.49949
0.00	0.50	0.30000
0.00	0.60	0.10051
0.00	0.70	0.10000
0.00	0.80	0.10000
0.00	0.90	0.10000
0.00	1.00	0.10000
0.10	0.00	1.00000
0.10	0.10	1.00000
0.10	0.20	0.99999
0.10	0.30	0.96215
0.10	0.40	0.50241
0.10	0.50	0.46324
0.10	0.60	0.10448
0.10	0.70	0.10041
0.10	0.80	0.10018
0.10	0.90	0.10011
0.10	1.00	0.10000

.
.
.

Table 8.8: (*Continued.*) Abbreviated numerical output for eqs. (8.4), `ncase=2`

```
Output for t = 0.2,...
     0.8 removed
          .
          .
          .
 1.00  0.00   1.00000
 1.00  0.10   1.00000
 1.00  0.20   1.00000
 1.00  0.30   1.00000
 1.00  0.40   1.00000
 1.00  0.50   1.00000
 1.00  0.60   1.00000
 1.00  0.70   1.00000
 1.00  0.80   1.00000
 1.00  0.90   0.69654
 1.00  1.00   0.10000


 ncall = 1296
```

This example also demonstrates (1) the effectiveness of the 5PBU FDs in resolving steep, moving fronts, and (2) the agreement of the resulting numerical solution with the analytical solution of eq. (8.4e) for the integer case $\alpha = 2$. In the limit $\nu \to 0$, eq. (8.4a) becomes the first order, nonlinear advection equation (entirely hyperbolic)

$$\frac{\partial u}{\partial t} = -u \frac{\partial u}{\partial x} \qquad (8.4f)$$

which also propagates steep moving fronts and discontinuities. So ν is a sensitive parameter and as it approaches zero, special approximations are required (to accommodate steep moving fronts and discontinuities).

The next example application is the Fokker-Planck (FP) equation, which in the one-sided form, is similar to eq. (8.4a).

8.4   FRACTIONAL FOKKER-PLANCK EQUATION

The FP equation is a convection-diffusion (CD) PDE that can be integrated numerically by the methods considered in Chapter 7. The two-sided version of the FP (eq. (7.1a) restated with $q_s(x,t) = 0$) is

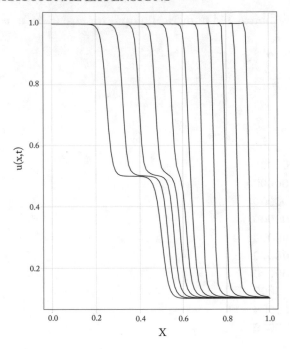

Figure 8.11: Numerical solution of eqs. (8.4), ncase=2.

$$\frac{\partial u(x,t)}{\partial t} = -v\frac{\partial u(x,t)}{\partial x} + d_+(x,t)\frac{\partial^\alpha u(x,t)}{\partial_+ x^\alpha} + d_-(x,t)\frac{\partial^\alpha u(x,t)}{\partial_- x^\alpha} \qquad (8.5a)$$

with $1 \leq \alpha \leq 2$.

The initial condition (IC) for eq. (8.5a) is

$$u(x, t = 0) = e^{-cx^2} \qquad (8.5b)$$

and homogeneous Dirichlet boundary conditions (BCs) are

$$u(x = x_l, t) = u(x = x_u, t) = 0 \qquad (8.5c,d)$$

For $d_-(x,t) = 0$, eq. (8.5a) is left-sided and for $d_+(x,t) = 0$, it is right-sided.

A main program for eqs. (8.5) follow.

8.4.1   MAIN PROGRAM

Listing 8.7: Main program for eqs. (8.5).

```
#
```

```
# Two-sided fractional Fokker-Planck
#
# du/dt=-v*du/dx+d+(x,t)d^alpha u/d+x^alpha
#                +d-(x,t)d^alpha u/d-x^alpha
#
# a < x < b, 0 < t < tf
#
# Delete previous workspaces
  rm(list=ls(all=TRUE))
#
# Access library, ODE/MOL routines
  library("deSolve");
  setwd("f:/fractional/sfpde/chap8/ex3");
  source("pde1a.R");
#
# Parameters
  ncase=1;
  if(ncase==1)alpha=2;
  if(ncase==2)alpha=1.1;
  v=1;c=2;
#
# Left diffusion coefficient
  d_left=function(x) 0.5;
#
# Right diffusion coefficient
  d_right=function(x) 0.5;
#
# IC
  f=function(x,t) exp(-c*x^2);
#
# BCs
  g_0=function(t) 0;
  g_L=function(t) 0;
#
# Grid in x
  xl=-5;xu=10;nx=101;
  dx=(xu-xl)/(nx-1);
  xj=seq(from=xl,to=xu,by=dx);
#
```

```
# Interval in t
  t0=0;tf=2;nt=6;
  dt=(tf-t0)/(nt-1);
  tout=seq(from=0,to=tf,by=dt);
#
# a_jk coefficients
  nx=nx-1;
  A=matrix(0,nrow=nx-1,ncol=nx);
  for(j in 1:(nx-1)){
    for(k in 0:j){
      if(k==0){
      A[j,k+1]=(j-1)^(3-alpha)-j^(2-alpha)*(j-3+alpha);
      } else if(1 <= k && k<=j-1){
      A[j,k+1]=(j-k+1)^(3-alpha)-2*(j-k)^(3-alpha)+(j-k-1)^(3-
          alpha);
      } else {
      A[j,k+1]=1;
      }
    }
  }
  A=(dx^(2-alpha)/gamma(4-alpha))*A;
#
# b_jk coefficients
  B=matrix(0,nrow=nx-1,ncol=nx);
  for(j in 1:(nx-1)){
    for(k in j:nx){
      if(k==j){
        B[j,k]=1;
      } else if(j+1 <= k && k<=nx-1){
        B[j,k]=(k-j+1)^(3-alpha)-2*(k-j)^(3-alpha)+(k-j-1)^(3-
          alpha);
      } else {
        B[j,k]=(nx-j-1)^(3-alpha)-(nx-j)^(2-alpha)*(nx-j+alpha-3)
          ;
      }
    }
  }
  B=(dx^(2-alpha)/gamma(4-alpha))*B;
#
```

```
# Initial condition
  y0=rep(0,nx-1);
  for(j in 1:(nx-1)){
    y0[j]=f(xj[j+1]);
  }
  ncall=0;
#
# ODE integration
  out=lsode(y=y0,times=tout,func=pde1a,
      rtol=1e-6,atol=1e-6,maxord=5);
  nrow(out)
  ncol(out)
#
# Allocate array for u(x,t)
  nx=nx+1;
  u=matrix(0,nt,nx);
#
# u(x,t), x ne xl,xu
  for(i in 1:nt){
  for(j in 2:(nx-1)){
    u[i,j]=out[i,j];
  }
  }
#
# Reset boundary values
  for(i in 1:nt){
    u[,1] =g_0(tout[i]);
    u[,nx]=g_L(tout[i]);
  }
#
# Tabular numerical solution
  cat(sprintf("\n\n    alpha = %4.2f\n",alpha));
  cat(sprintf("\n      t     x     u(x,t)"));
  for(i in 1:nt){
  iv=seq(from=1,to=nx,by=10);
  for(j in iv){
    cat(sprintf("\n %6.2f%6.2f%10.5f",
      tout[i],xj[j],u[i,j]));
  }
```

```
  cat(sprintf("\n"));
  }
#
# Plot numerical solution
  matplot(xj,t(u),type="l",lwd=2,col="black",lty=1,
    xlab="x",ylab="u(x,t)",main="");
#
# Calls to ODE routine
  cat(sprintf("\n   ncall = %3d\n",ncall));
```

We can note the following details about Listing 8.7.

- Brief documentation comments are followed by deletion of previous files.

```
  # Two-sided fractional Fokker-Planck
  #
  # du/dt=-v*du/dx+d+(x,t)d^alpha u/d+x^alpha
  #                 +d-(x,t)d^alpha u/d-x^alpha
  #
  # a < x < b, 0 < t < tf
  #
  # Delete previous workspaces
    rm(list=ls(all=TRUE))
```

- The ODE integrator library deSolve is accessed. The ODE/MOL routine is pde1a.

```
  #
  # Access library, ODE/MOL routines
    library("deSolve");
    setwd("f:/fractional/sfpde/chap8/ex3");
    source("pde1a.R");
```

- Two cases are programmed. For ncase=1, $\alpha = 2$ and the integer form of eq. (8.5a) is used. For ncase=2, $\alpha = 1.1$, the fractional derivatives are shifted toward first order so that they add convection right to left in addition to the left to right convection of $-v\dfrac{\partial u}{\partial x}$. The convection terms therefore partially offset each other as observed in the output for ncase=2 (discussed subsequently).

```
  #
  # Parameters
    ncase=1;
```

```
if(ncase==1)alpha=2;
if(ncase==2)alpha=1.1;
v=1;c=2;
```

The velocity v in eq. (8.5a) and the constant in the Gaussian IC (8.5b), c, are also defined numerically.

- The left and right diffusion coefficients in eq. (8.5a) are defined.

```
#
# Left diffusion coefficient
  d_left=function(x) 0.5;
#
# Right diffusion coefficient
  d_right=function(x) 0.5;
```

- A function for the Gaussian IC (8.5b) (centered on $x = 0$) is defined.

```
#
# IC
  f=function(x,t) exp(-c*x^2);
```

- Functions for the homogeneous Dirchlet BCs (8.5c,d) are defined.

```
#
# BCs
  g_0=function(t) 0;
  g_L=function(t) 0;
```

- A spatial grid of 101 points is defined for the interval $x_l = -5 \le x \le x_u = 10$, so that grid values are $x = -5, -4.85, ..., 10$.

```
#
# Grid in x
  xl=-5;xu=10;nx=101;
  dx=(xu-xl)/(nx-1);
  xj=seq(from=xl,to=xu,by=dx);
```

- The interval in t, $t = t_0 = 0 \le t \le t = t_f = 2$, is defined with 6 points, so the output values are $t = 0, 0.4, ..., 2$.

```
#
# Interval in t
  t0=0;tf=2;nt=6;
  dt=(tf-t0)/(nt-1);
  tout=seq(from=0,to=tf,by=dt);
```

- The coefficient matrix A of eq. (1.2g) and the matrix B of eq. (7.4a) are defined.

- IC (8.5b) is implemented.

```
#
# Initial condition
  y0=rep(0,nx-1);
  for(j in 1:(nx-1)){
    y0[j]=f(xj[j+1]);
  }
  ncall=0;
```

- The 99 ODEs at the interior points in x are integrated with lsode.

```
#
# ODE integration
  out=lsode(y=y0,times=tout,func=pde1a,
      rtol=1e-6,atol=1e-6,maxord=5);
  nrow(out)
  ncol(out)
```

The ODE/MOL routine pde1a is considered next. The solution matrix out has dimensions $6 \times 99 + 1 = 100$ as confirmed in the subsequent output (the offset $+1$ provides for the value of t for each of the 6 derivative vectors).

- The solution out returned by lsode is placed in a matrix u as a function of x and t.

```
#
# Allocate array for u(x,t)
  nx=nx+1;
  u=matrix(0,nt,nx);
#
# u(x,t), x ne xl,xu
  for(i in 1:nt){
  for(j in 2:(nx-1)){
```

```
    u[i,j]=out[i,j];
  }
}
```

- The boundary values of the solution are defined as a function of t (since these values are not produced from the integration of an ODE, they are not returned by lsode).

```
#
# Reset boundary values
  for(i in 1:nt){
    u[,1] =g_0(tout[i]);
    u[,nx]=g_L(tout[i]);
  }
```

In this case, the boundary values are given by BCs (8.5c,d).

- The numerical solution is displayed for every tenth value of x from by=10.

```
#
# Tabular numerical solution
  cat(sprintf("\n\n   alpha = %4.2f\n",alpha));
  cat(sprintf("\n      t     x     u(x,t)"));
  for(i in 1:nt){
  iv=seq(from=1,to=nx,by=10);
  for(j in iv){
    cat(sprintf("\n %6.2f%6.2f%10.5f",
      tout[i],xj[j],u[i,j]));
  }
  cat(sprintf("\n"));
  }
```

- The numerical solution is plotted. The transpose t(u) is required so that xj and u are comformable (dimensionally consistent). The solution is plotted parametrically in t.

```
#
# Plot numerical solution
  matplot(xj,t(u),type="l",lwd=2,col="black",lty=1,
    xlab="x",ylab="u(x,t)",main="");
```

- The number of calls to pde1a is displayed at end of the solution.

```
#
# Calls to ODE routine
  cat(sprintf("\n   ncall = %3d\n",ncall));
```

8.4.2   ODE/MOL ROUTINE

The ODE/MOL routine (8.5a) is the same as in Listing 8.6 except that the programming of eq. (8.4a) is changed to eq. (8.5a).

Listing 8.6

```
ut[j]=cd*vis*ut[j]-u[j]*ux[j];
```

Current pde1a

```
ut[j]=-v*ux[j]+frac_diff;
```

The numerical output from the preceding main program of Listing 8.7 and the ODE/-MOL routine of Listing 8.6, with the change in the PDE noted above, is considered next.

8.4.3   MODEL OUTPUT

Abbreviated numerical output for ncase=1, $\alpha = 2$ is shown in Table 8.9.

We can note the following details about the output in Table 8.9.

- The output matrix out from lsode is $6 \times 99 + 1 = 100$ as explained previously.

    [1] 6

    [1] 100

- The interval in x is $-5 \leq x \leq 10$ as programmed in Listing 8.7 and the output interval in x is 1.5 (with by=10).

- The interval in t is $0 \leq t \leq 2$ with the output interval 0.4 as programmed in Listing 8.6.

- The computational effort is modest.

    ncall = 1074

Figure 8.12 indicates the Gaussian IC is dispersed by the two fractional derivative terms in eq. (8.5a) (which are integer with $\alpha = 2$, i.e., just Fickian or parabolic diffusion) and moves left to right according to the convection term $-v\dfrac{\partial u}{\partial x}$.

Abbreviated output for ncase=2 ($\alpha = 1.1$) is shown in Table 8.10.

We note by comparing Figs. 8.12 and 8.13 that by decreasing α from 2 to 1.1, the fractional derivatives in eq. (8.5a) are shifted toward first order so that they add convection right to left in

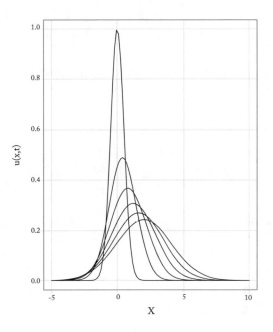

Figure 8.12: Numerical solution of eqs. (8.5), ncase=1.

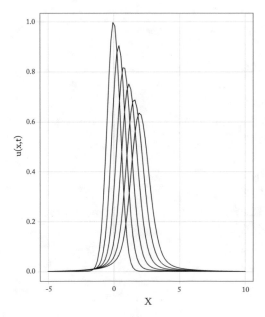

Figure 8.13: Numerical solution of eqs. (8.5), ncase=2.

Table 8.9: Abbreviated numerical output for eqs. (8.5), ncase=1 (*Continues.*)

[1] 6

[1] 100

```
  alpha = 2.00

     t      x    u(x,t)
  0.00 -5.00   0.00000
  0.00 -3.50   0.00000
  0.00 -2.00   0.00034
  0.00 -0.50   0.60653
  0.00  1.00   0.13534
  0.00  2.50   0.00000
  0.00  4.00   0.00000
  0.00  5.50   0.00000
  0.00  7.00   0.00000
  0.00  8.50   0.00000
  0.00 10.00   0.00000

  0.40 -5.00   0.00000
  0.40 -3.50   0.00036
  0.40 -2.00   0.03117
  0.40 -0.50   0.33246
  0.40  1.00   0.41047
  0.40  2.50   0.05978
  0.40  4.00   0.00112
  0.40  5.50   0.00000
  0.40  7.00   0.00000
  0.40  8.50   0.00000
  0.40 10.00   0.00000
```

Table 8.9: (*Continued.*) Abbreviated numerical output for eqs. (8.5), ncase=1

.
.
.

```
Output for t = 0.8,...,
     1.6 removed
```

.
.
.

```
2.00 -5.00    0.00000
2.00 -3.50    0.00664
2.00 -2.00    0.03680
2.00 -0.50    0.11635
2.00  1.00    0.21594
2.00  2.50    0.23549
2.00  4.00    0.15118
2.00  5.50    0.05732
2.00  7.00    0.01289
2.00  8.50    0.00173
2.00 10.00    0.00000

ncall = 1074
```

addition to the left to right convection of $-v\dfrac{\partial u}{\partial x}$. The convection terms therefore partially offset each other as observed in Fig. 8.13 (the pulses are compressed in x).

This completes the discussion of the two-sided Fokker Planck eq. (8.5a).

Table 8.10: Abbreviated numerical output for eqs. (8.5), ncase=2 (*Continues.*)

[1] 6

[1] 100

```
alpha = 1.10

    t      x      u(x,t)
 0.00 -5.00   0.00000
 0.00 -3.50   0.00000
 0.00 -2.00   0.00034
 0.00 -0.50   0.60653
 0.00  1.00   0.13534
 0.00  2.50   0.00000
 0.00  4.00   0.00000
 0.00  5.50   0.00000
 0.00  7.00   0.00000
 0.00  8.50   0.00000
 0.00 10.00   0.00000

 0.40 -5.00   0.00000
 0.40 -3.50   0.00131
 0.40 -2.00   0.00469
 0.40 -0.50   0.22243
 0.40  1.00   0.46910
 0.40  2.50   0.00727
 0.40  4.00   0.00167
 0.40  5.50   0.00065
 0.40  7.00   0.00028
 0.40  8.50   0.00011
 0.40 10.00   0.00000
```

Table 8.10: (*Continued.*) Abbreviated numerical output for eqs. (8.5), `ncase=2`

```
        .
        .
        .

Output for t = 0.8,...,
     1.6 removed
        .
        .
        .
2.00 -5.00    0.00000
2.00 -3.50    0.00173
2.00 -2.00    0.00601
2.00 -0.50    0.02065
2.00  1.00    0.22536
2.00  2.50    0.45551
2.00  4.00    0.04067
2.00  5.50    0.00885
2.00  7.00    0.00318
2.00  8.50    0.00113
2.00 10.00    0.00000

ncall = 651
```

8.5   FRACTIONAL BURGERS-HUXLEY EQUATION

As the next example application, we consider the fractional Burgers-Huxley (BH) equation

$$\frac{\partial u}{\partial t} + u^2 \frac{\partial u}{\partial x} - \frac{\partial^\alpha u}{\partial x^\alpha} = \frac{2}{3}u^3(1 - u^2) \tag{8.6a}$$

The initial condition (IC) for eq. (8.6a) is based on the analytical solution (discussed below).

$$u(x, t = 0) = \left[\frac{1}{2} + \frac{1}{2}\tanh\left(\frac{1}{3}x\right)\right]^{\frac{1}{2}} \tag{8.6b}$$

The BCs are:
Homogeneous Dirichlet:

$$u(x = x_l, t) = u(x = x_u, t) = 0 \tag{8.6c,d}$$

Homogeneous Neumann:

$$\frac{\partial u(x = x_l, t)}{\partial x} = \frac{\partial u(x = x_u, t)}{\partial x} = 0 \tag{8.6e,f}$$

Homogeneous Robin:

$$c_2 \frac{\partial u(x = x_l, t)}{\partial x} + c_1 u(x = x_l, t) = 0 \tag{8.6g}$$

$$c_4 \frac{\partial u(x = x_u, t)}{\partial x} + c_4 u(x = x_u, t) = 0 \tag{8.6h}$$

The analytical solution for $\alpha = 2$ is

$$u_a(x, t) = \left[\frac{1}{2} + \frac{1}{2} \tanh\left(\frac{1}{9}(3x + t) \right) \right]^{\frac{1}{2}} \tag{8.6i}$$

$u_a(x, t)$ is a traveling wave solution with Lagrangian variable $k(x - vt) = \frac{1}{9}(3x + t) = \frac{1}{3}(x - (-1/3)t)$ or $v = -1/3$ so the solution travels right to left.

8.5.1 MAIN PROGRAM

A main program for eqs. (8.6) follows.

Listing 8.8: Main program for eqs. (8.6).

```
#
# Fractional Burgers-Huxley
#
#    ut=-u^2*ux+(d^alpha u/dx^alpha)+(2/3)*u^3*(1-u^2)
#
#    xl < x < xu, 0 < t < tf, xl=-20, xu=10
#
#    u(x,t=0)=ua(x,t=0)
#
#    Dirichlet, Neumann, Robin BCs
#
#    ua(x,t) (analytical solution is
#            multiline function)
#
# Delete previous workspaces
  rm(list=ls(all=TRUE))
#
# Access functions for numerical solution
```

```
  library("deSolve");
  setwd("f:/fractional/sfpde/chap8/ex4");
  source("pde1a.R");source("ua.R");
#
# Parameters
  alpha=2;
# alpha=1.5;
  ncase=1;
#
# Dirichlet BCs
  if(ncase==1){
#
#   Boundary condition functions
    g_0=function(t) 0;
    g_L=function(t) 1;
#
#   Boundary condition coefficients
    c_1=function(t) 1;
    c_2=function(t) 0;
    c_3=function(t) 1;
    c_4=function(t) 0;
  }
#
# Neumann BCs
  if(ncase==2){
#
#   Boundary condition functions
    g_0=function(t) 0;
    g_L=function(t) 0;
#
#   Boundary condition coefficients
    c_1=function(t) 0;
    c_2=function(t) 1;
    c_3=function(t) 0;
    c_4=function(t) 1;
  }
#
# Robin BCs
  if(ncase==3){
```

```
#
#    Boundary condition functions
     g_0=function(t) 0;
     g_L=function(t) 1;
#
#    Boundary condition coefficients
     c_1=function(t) -1;
     c_2=function(t)  1;
     c_3=function(t)  1;
     c_4=function(t)  1;
   }
#
# Spatial grid
  xl=-20;xu=10;nx=51;dx=(xu-xl)/(nx-1);
  xj=seq(from=xl,to=xu,by=dx);
  cd=dx^(-alpha)/gamma(4-alpha);
  r12dx=1/(12*dx);
#
# Independent variable for ODE integration
  t0=0;tf=15;nt=4;dt=(tf-t0)/(nt-1);
  tout=seq(from=t0,to=tf,by=dt);
#
# a_jk coefficients
  A=matrix(0,nrow=nx-2,ncol=nx-1);
  for(j in 1:(nx-2)){
    for(k in 0:j){
    if (k==0){
      A[j,k+1]=(j-1)^(3-alpha)-j^(2-alpha)*(j-3+alpha);
    } else if (1 <= k && k<=j-1){
      A[j,k+1]=(j-k+1)^(3-alpha)-2*(j-k)^(3-alpha)+(j-k-1)^(3-
         alpha);
    } else
      A[j,k+1]=1;
    }
  }
#
# Initial condition
  u0=rep(0,nx-2);
  for(j in 1:(nx-2)){
```

```
    u0[j]=ua(xj[j+1],0);}
  ncall=0;
#
# ODE integration
  out=lsodes(func=pde1a,y=u0,times=tout,
             sparsetype="sparseint")
  nrow(out)
  ncol(out)
#
# Allocate array for u(x,t)
  u=matrix(0,nt,nx);
#
# u(x,t), x ne xl,xu
  for(i in 1:nt){
    for(j in 2:(nx-1)){
      u[i,j]=out[i,j];
    }
  }
#
# Reset boundary values
  for(i in 1:nt){
      u[i,1]=0;
    u[i,nx]=1;
  }
#
# Display parameters
  cat(sprintf("\n   alpha = %4.2f\n",alpha));
#
# ncase=1,2
  if((ncase==1)|(ncase==2)){
#
# Numerical, analytical solutions, difference
  uap=matrix(0,nt,nx);
  for(i in 1:nt){
    for(j in 1:nx){
      uap[i,j]=ua(xj[j],tout[i]);
    }
  max_err=max(abs(u-uap));
  }
```

```
#
# Tabular numerical, analytical solutions,
# difference
  cat(sprintf("\n        t        x      u(x,t)     ua(x,t)              diff
     "));
  for(i in 1:nt){
  iv=seq(from=1,to=nx,by=5);
  for(j in iv){
    cat(sprintf("\n %6.2f%8.2f%10.5f%10.5f%12.3e",
      tout[i],xj[j],u[i,j],uap[i,j],u[i,j]-uap[i,j]));
  }
  cat(sprintf("\n"));
  }
#
# Plot numerical, analytical solutions
  matplot(xj,t(u),type="l",lwd=2,col="black",lty=1,
    xlab="x",ylab="u(x,t)",main="");
  matpoints(xj,t(uap),pch="o",col="black");
  }
#
# ncase=3
  if(ncase>2){
#
# Tabular numerical solution
  cat(sprintf("\n        t        x      u(x,t)"));
  for(i in 1:nt){
  iv=seq(from=1,to=nx,by=5);
  for(j in iv){
    cat(sprintf("\n %6.2f%8.2f%10.5f",
      tout[i],xj[j],u[i,j]));
  }
  cat(sprintf("\n"));
  }
#
# Plot numerical solution
  matplot(xj,t(u),type="l",lwd=2,col="black",lty=1,
    xlab="x",ylab="u(x,t)",main="");
  }
#
```

```
# Calls to ODE routine
  cat(sprintf("\n\n  ncall = %3d\n",ncall));
```

- Brief documentation comments are followed by deletion of previous files.

```
#
# Fractional Burgers-Huxley
#
#   ut=-u^2*ux+(d^alpha u/dx^alpha)+(2/3)*u^3*(1-u^2)
#
#   xl < x < xu, 0 < t < tf, xl=-20, xu=10
#
#   u(x,t=0)=ua(x,t=0)
#
#   Dirichlet, Neumann, Robin BCs
#
#   ua(x,t) (analytical solution is
#               multiline function)
#
# Delete previous workspaces
  rm(list=ls(all=TRUE))
```

- The ODE integrator library deSolve is accessed. The ODE/MOL routine is pde1a, discussed subsequently. ua is a function for analytical solution (8.6i).

```
#
# Access functions for numerical solution
  library("deSolve");
  setwd("f:/fractional/sfpde/chap8/ex4");
  source("pde1a.R");source("ua.R");
```

- The parameters are defined numerically. Initially $\alpha = 2$ so eq. (8.6a) is integer with the analytical solution of eq. (8.6i).

```
#
# Parameters
  alpha=2;
# alpha=1.5;
  ncase=1;
```

ncase=1 corresponds to homogeneous Dirichlet BCs (8.6c,d).

- For ncase=1, the functions and coefficients in eqs. (3.4a,b) are defined for Dirichlet BCs (8.6c,d).

```
#
# Dirichlet BCs
  if(ncase==1){
#
#    Boundary condition functions
     g_0=function(t) 0;
     g_L=function(t) 1;
#
#    Boundary condition coefficients
     c_1=function(t) 1;
     c_2=function(t) 0;
     c_3=function(t) 1;
     c_4=function(t) 0;
  }
```

- For ncase=2, the functions and coefficients in eqs. (3.4a,b) are defined for Neumann BCs (8.6e,f).

```
#
# Neumann BCs
  if(ncase==2){
#
#    Boundary condition functions
     g_0=function(t) 0;
     g_L=function(t) 0;
#
#    Boundary condition coefficients
     c_1=function(t) 0;
     c_2=function(t) 1;
     c_3=function(t) 0;
     c_4=function(t) 1;
  }
```

- For ncase=3, the functions and coefficients in eqs. (3.4a,b) are defined for Robin BCs (8.6g,h).

```
#
# Robin BCs
```

```
   if(ncase==3){
#
#    Boundary condition functions
     g_0=function(t) 0;
     g_L=function(t) 1;
#
#    Boundary condition coefficients
     c_1=function(t) -1;
     c_2=function(t)  1;
     c_3=function(t)  1;
     c_4=function(t)  1;
   }
```

- A spatial grid of 51 points is defined for the interval $x_l = -20 \le x \le x_u = 10$, so that grid values are incremented by $30/50 = 0.6$.

```
#
# Spatial grid
  xl=-20;xu=10;nx=51;dx=(xu-xl)/(nx-1);
  xj=seq(from=xl,to=xu,by=dx);
  cd=dx^(-alpha)/gamma(4-alpha);
  r12dx=1/(12*dx);
```

- The interval in t, $t = t_0 = 0 \le t \le t = t_f = 15$, is defined with 4 points, so the output values are $t = 0, 5, 10, 15$.

```
#
# Independent variable for ODE integration
  t0=0;tf=15;nt=4;dt=(tf-t0)/(nt-1);
  tout=seq(from=t0,to=tf,by=dt);
```

- The A coefficients of eq. (1.2g) are computed.

- IC (8.6b) is defined with the analytical solution ua.

```
#
# Initial condition
  u0=rep(0,nx-2);
  for(j in 1:(nx-2)){
    u0[j]=ua(xj[j+1],0);}
  ncall=0;
```

- The ODE at the interior points in x are integrated by lsodes.

```
#
# ODE integration
  out=lsodes(func=pde1a,y=u0,times=tout,
             sparsetype="sparseint")
  nrow(out)
  ncol(out)
```

The solution matrix out has the dimensions $6 \times 101 - 2 + 1 = 100$ as demonstrated in the numerical output that follows. The offset of $+1$ indicates that the value of t is included as the first elements of each solution vector of 99 ODE solutions.

- The numerical solution from out (returned by lsodes) is placed in array u (with dimensions u(6,101)).

```
#
# Allocate array for u(x,t)
  u=matrix(0,nt,nx);
#
# u(x,t), x ne xl,xu
  for(i in 1:nt){
    for(j in 2:(nx-1)){
      u[i,j]=out[i,j];
    }
  }
```

- The boundary values at $x = -20, x = 10$ are reset.

```
#
# Reset boundary values
  for(i in 1:nt){
      u[i,1]=0;
      u[i,nx]=1;
  }
```

These boundary values ($u(x = x_l = -20, t) = 0, u(x = x_u = 10, t) = 1$) reflect prior knowledge of the solution at the boundaries. These values are not returned by lsodes since they are not set by the integration of ODEs.

- The order of the fractional derivatives in eq. (8.6a), α, is displayed.

```
#
# Display parameters
  cat(sprintf("\n   alpha = %4.2f\n",alpha));
```

- For ncase=1,2 analytical solution (8.6i) is placed in the array uap and the maximum difference between the numerical and analytical solutions is placed in max_err (the R utilities max, abs operate on arrays).

```
#
# ncase=1,2
  if((ncase==1)|(ncase==2)){
#
# Numerical, analytical solutions, difference
  uap=matrix(0,nt,nx);
  for(i in 1:nt){
    for(j in 1:nx){
      uap[i,j]=ua(xj[j],tout[i]);
    }
  max_err=max(abs(u-uap));
  }
#
# Tabular numerical, analytical solutions,
# difference
  cat(sprintf("\n       t       x      u(x,t)   ua(x,t)        diff"));
  for(i in 1:nt){
  iv=seq(from=1,to=nx,by=10);
  for(j in iv){
    cat(sprintf("\n %6.2f%8.2f%10.5f%10.5f%12.3e",
      tout[i],xj[j],u[i,j],uap[i,j],u[i,j]-uap[i,j]));
  }
  cat(sprintf("\n"));
  }
```

The numerical and analytical solutions and the difference are displayed for every tenth value of x (from by=10). Since the interval in x is $(10 - (-20))/50 = 0.6$, the output interval in x is $10(0.6) = 6$.

- The numerical solution is plotted with maplot and the analytical solutions is superimposed with matpoints. The transposes t(u), t(uap) are required so that u, uap are conformable (same dimension) with xj. The solutions are plotted parametrically in t (6 solution curves).

```
  #
  # Plot numerical, analytical solutions
    matplot(xj,t(u),type="l",lwd=2,col="black",lty=1,
      xlab="x",ylab="u(x,t)",main="");
    matpoints(xj,t(uap),pch="o",col="black");
    }
```

- For ncase=3, an analytical solution is not used (not readily available).

```
  #
  # ncase=3
    if(ncase>2){
  #
  # Tabular numerical solution
    cat(sprintf("\n       t       x    u(x,t)"));
    for(i in 1:nt){
    iv=seq(from=1,to=nx,by=5);
    for(j in iv){
      cat(sprintf("\n %6.2f%8.2f%10.5f",
        tout[i],xj[j],u[i,j]));
    }
    cat(sprintf("\n"));
    }
```

- The numerical solution is plotted.

```
  #
  # Plot numerical solution
    matplot(xj,t(u),type="l",lwd=2,col="black",lty=1,
      xlab="x",ylab="u(x,t)",main="");
```

- The number of calls to pde1a is displayed at the end of the solution.

```
  #
  # Calls to ODE routine
    cat(sprintf("\n\n  ncall = %3d\n",ncall));
```

8.5.2   ODE/MOL ROUTINE

The ODE/MOL routine pde1a called by lsodes is similar to the routines in Listings 8.2 and 8.6 and therefore is not listed here. The programming of eq. (8.6a) is the essential difference.

```
ut[j]=-(u[j]^2)*ux[j]+cd*ut[j]+(2/3)*(u[j]^3)*(1-u[j]^2);
```

The programming of the nonlinear convective term $u^2 \dfrac{\partial u}{\partial x}$ and the nonlinear source term $\dfrac{2}{3} u^3 (1 - u^2)$ is noteworthy.

The programming of the analytical solution of eq. (8.6i) is straightforward and does not require further discussion.

```
  ua=function(x,t){
#
# Function ua computes the exact solution of the
# Burgers-Huxley equation for comparison with the
# numerical solution
#
# Analytical solution
  expp=exp( (1/3)*x+(1/9)*t);
  expm=exp(-(1/3)*x-(1/9)*t);
  ua=((1/2)*(1+(expp-expm)/(expp+expm)))^0.5;
#
# Return solution
  return(c(ua));
  }
```

8.5.3   MODEL OUTPUT

The numerical solution for eqs. (8.6) (ncase=1, $\alpha = 2$, Dirichlet BCs) is shown in Table 8.11. The agreement between the numerical and analytical solutions is indicated in Fig. 8.14.

The traveling wave solution of eq. (8.6i) is apparent with movement right to left since the velocity is $v = -1/3$ as discussed previously.

The numerical solution for (ncase=2, $\alpha = 2$, Neumann BCs) is shown in Table 8.12. The agreement between the numerical and analytical solutions is clear. The graphical output is essentially the same as in Fig. 8.14 and is not repeated here.

The numerical solution for (ncase=3, $\alpha = 2$, Robin BCs) is shown in Table 8.13.

The graphical output is in Fig. 8.15.

Table 8.11: Numerical solution for eqs. (8.6), ncase=1, $\alpha = 2$

[1] 4

[1] 50

alpha = 2.00

t	x	u(x,t)	ua(x,t)	diff
0.00	-20.00	0.00000	0.00127	-1.273e-03
0.00	-14.00	0.00940	0.00940	0.000e+00
0.00	-8.00	0.06932	0.06932	0.000e+00
0.00	-2.00	0.45674	0.45674	0.000e+00
0.00	4.00	0.96697	0.96697	0.000e+00
0.00	10.00	1.00000	0.99936	6.357e-04
5.00	-20.00	0.00000	0.00222	-2.218e-03
5.00	-14.00	0.01633	0.01639	-5.949e-05
5.00	-8.00	0.12022	0.12022	-7.776e-06
5.00	-2.00	0.66638	0.66684	-4.565e-04
5.00	4.00	0.98875	0.98876	-1.045e-05
5.00	10.00	1.00000	0.99979	2.094e-04
10.00	-20.00	0.00000	0.00387	-3.866e-03
10.00	-14.00	0.02826	0.02855	-2.953e-04
10.00	-8.00	0.20606	0.20652	-4.607e-04
10.00	-2.00	0.84150	0.84182	-3.157e-04
10.00	4.00	0.99623	0.99626	-2.797e-05
10.00	10.00	1.00000	0.99993	6.895e-05
15.00	-20.00	0.00000	0.00674	-6.738e-03
15.00	-14.00	0.04898	0.04973	-7.405e-04
15.00	-8.00	0.34377	0.34526	-1.490e-03
15.00	-2.00	0.93828	0.93851	-2.288e-04
15.00	4.00	0.99875	0.99876	-1.779e-05
15.00	10.00	1.00000	0.99998	2.270e-05

ncall = 112

Table 8.12: Numerical solution for eqs. (8.6), ncase=2, $\alpha = 2$

[1] 4

[1] 50

alpha = 2.00

t	x	u(x,t)	ua(x,t)	diff
0.00	-20.00	0.00000	0.00127	-1.273e-03
0.00	-14.00	0.00940	0.00940	0.000e+00
0.00	-8.00	0.06932	0.06932	0.000e+00
0.00	-2.00	0.45674	0.45674	0.000e+00
0.00	4.00	0.96697	0.96697	0.000e+00
0.00	10.00	1.00000	0.99936	6.357e-04
5.00	-20.00	0.00000	0.00222	-2.218e-03
5.00	-14.00	0.01645	0.01639	6.546e-05
5.00	-8.00	0.12022	0.12022	-7.594e-06
5.00	-2.00	0.66638	0.66684	-4.568e-04
5.00	4.00	0.98875	0.98876	-1.007e-05
5.00	10.00	1.00000	0.99979	2.094e-04
10.00	-20.00	0.00000	0.00387	-3.866e-03
10.00	-14.00	0.02885	0.02855	2.994e-04
10.00	-8.00	0.20608	0.20652	-4.416e-04
10.00	-2.00	0.84151	0.84182	-3.137e-04
10.00	4.00	0.99623	0.99626	-2.779e-05
10.00	10.00	1.00000	0.99993	6.895e-05
15.00	-20.00	0.00000	0.00674	-6.738e-03
15.00	-14.00	0.05041	0.04973	6.874e-04
15.00	-8.00	0.34390	0.34526	-1.362e-03
15.00	-2.00	0.93828	0.93851	-2.229e-04
15.00	4.00	0.99875	0.99876	-1.771e-05
15.00	10.00	1.00000	0.99998	2.270e-05

ncall = 104

Table 8.13: Numerical solution for eqs. (8.6), ncase=3, $\alpha = 2$

[1] 4

[1] 50

```
  alpha = 2.00

      t        x     u(x,t)
   0.00   -20.00   0.00000
   0.00   -14.00   0.00940
   0.00    -8.00   0.06932
   0.00    -2.00   0.45674
   0.00     4.00   0.96697
   0.00    10.00   1.00000

   5.00   -20.00   0.00000
   5.00   -14.00   0.01638
   5.00    -8.00   0.12022
   5.00    -2.00   0.66638
   5.00     4.00   0.98875
   5.00    10.00   1.00000

  10.00   -20.00   0.00000
  10.00   -14.00   0.02846
  10.00    -8.00   0.20607
  10.00    -2.00   0.84150
  10.00     4.00   0.99623
  10.00    10.00   1.00000

  15.00   -20.00   0.00000
  15.00   -14.00   0.04940
  15.00    -8.00   0.34381
  15.00    -2.00   0.93828
  15.00     4.00   0.99875
  15.00    10.00   1.00000

  ncall = 106
```

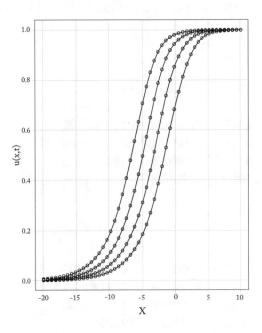

Figure 8.14: Numerical solution of eqs. (8.6), ncase=1, $\alpha = 2$.

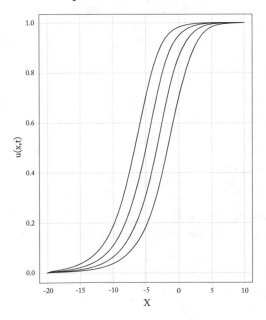

Figure 8.15: Numerical solution of eqs. (8.6), ncase=3, $\alpha = 2$.

A comparison of the solutions for ncase=1,2,3 at $t = 15$ shows the effect of the variations in the BCs is small.

ncase=1

```
15.00   -20.00    0.00000    0.00674   -6.738e-03
15.00   -14.00    0.04898    0.04973   -7.405e-04
15.00    -8.00    0.34377    0.34526   -1.490e-03
15.00    -2.00    0.93828    0.93851   -2.288e-04
15.00     4.00    0.99875    0.99876   -1.779e-05
15.00    10.00    1.00000    0.99998    2.270e-05
```

ncase=2

```
15.00   -20.00    0.00000    0.00674   -6.738e-03
15.00   -14.00    0.05041    0.04973    6.874e-04
15.00    -8.00    0.34390    0.34526   -1.362e-03
15.00    -2.00    0.93828    0.93851   -2.229e-04
15.00     4.00    0.99875    0.99876   -1.771e-05
15.00    10.00    1.00000    0.99998    2.270e-05
```

ncase=3

```
15.00   -20.00    0.00000
15.00   -14.00    0.04940
15.00    -8.00    0.34381
15.00    -2.00    0.93828
15.00     4.00    0.99875
15.00    10.00    1.00000
```

A principal use of the preceding routines is to demonstrate the effect of the order of the fractional derivative α in eq. (8.6a). Here we consider changing α from 2 to 1.5 using

```
#
# Parameters
  alpha=2;
  alpha=1.5;
```

in Listing 8.8. Also, since ncase=1,2,3 does not include an analytical solution, the output at the end of Listing 8.8 is changed to

```
#
# ncase=1,2,3
#
# Tabular numerical solution
  cat(sprintf("\n      t        x     u(x,t)"));
  for(i in 1:nt){
  iv=seq(from=1,to=nx,by=5);
  for(j in iv){
    cat(sprintf("\n %6.2f%8.2f%10.5f",
      tout[i],xj[j],u[i,j]));
  }
  cat(sprintf("\n"));
  }
#
# Plot numerical solution
  matplot(xj,t(u),type="l",lwd=2,col="black",lty=1,
    xlab="x",ylab="u(x,t)",main="");
```

Since decreasing α from 2 to 1.5 moves the fractional derivative in eq. (8.6a) from parabolic (second order, diffusive) closer to hyperbolic (first order, convective), the convection of the solution increases from right to left. In fact, the solution reaches the left boundary at $x_l = -20$ and therefore departs from zero. In order to accommodate this movement, the left boundary is extended from $x_l = -20$ to $x_l = -25$.

```
#
# Spatial grid
  xl=-25;xu=10;nx=51;dx=(xu-xl)/(nx-1);
```

With nx=51, the grid spacing is now $(10 - (-25))/50 = 0.7$ as reflected in the output that follows.

The output for the three cases (the three types of BCs) is considered next. For ncase=1 ($\alpha = 1.5$), the solution is in Table 8.14. The graphical output is in Fig. 8.16.

The increased movement to the left is evident from comparing Figs. 8.14 and 8.16 (note the extension of the left boundary from -20 to 25 and the spacing in x of $(10)0.7 = 7$).

Table 8.15 indicates that variation in the BCs with ncase=1,2,3 has a minor effect on the solution (as with $\alpha = 2$) so that the graphical output for ncase=2,3 is not included here.

This concludes the discussion of the BH equation (8.6a). This chapter is concluded with a final example application, the Fitzhugh-Nagumo equation.

Table 8.14: Numerical solution for eqs. (8.6), ncase=1, $\alpha = 1.5$

[1] 4

[1] 50

```
  0.00   -25.00     0.00000
  0.00   -18.00     0.00248
  0.00   -11.00     0.02555
  0.00    -4.00     0.25489
  0.00     3.00     0.93851
  0.00    10.00     1.00000

  5.00   -25.00     0.00000
  5.00   -18.00     0.00625
  5.00   -11.00     0.06754
  5.00    -4.00     0.60511
  5.00     3.00     0.97834
  5.00    10.00     1.00000

 10.00   -25.00     0.00000
 10.00   -18.00     0.01648
 10.00   -11.00     0.18713
 10.00    -4.00     0.89670
 10.00     3.00     0.99037
 10.00    10.00     1.00000

 15.00   -25.00     0.00000
 15.00   -18.00     0.04517
 15.00   -11.00     0.49217
 15.00    -4.00     0.97135
 15.00     3.00     0.99426
 15.00    10.00     1.00000

ncall = 186
```

Table 8.15: Numerical solutions for eqs. (8.6), ncase=1,2,3, $\alpha = 1.5$, $t = 15$

```
ncase=1, alpha=1.5

15.00   -25.00     0.00000
15.00   -18.00     0.04517
15.00   -11.00     0.49217
15.00    -4.00     0.97135
15.00     3.00     0.99426
15.00    10.00     1.00000


ncase=2, alpha=1.5

15.00   -25.00     0.00000
15.00   -18.00     0.04813
15.00   -11.00     0.49528
15.00    -4.00     0.97177
15.00     3.00     0.99447
15.00    10.00     1.00000


ncase=3, alpha=1.5

15.00   -25.00     0.00000
15.00   -18.00     0.04584
15.00   -11.00     0.49289
15.00    -4.00     0.97144
15.00     3.00     0.99431
15.00    10.00     1.00000
```

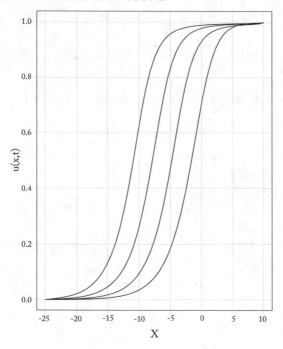

Figure 8.16: Numerical solution of eqs. (8.6), `ncase=1`, $\alpha = 1.5$.

8.6   FRACTIONAL FITZHUGH-NAGUMO EQUATION

The fractional Fizthugh-Nagumo (FN) equation is

$$\frac{\partial u}{\partial t} = D \frac{\partial^\alpha u}{\partial x^\alpha} + u(u-1)(a-u) \tag{8.7a}$$

with arbitraty constants a, D.

Equation (8.7a) is first order in t therefore requires one initial condition (IC).

$$u(x, t=0) = g_1(x) \tag{8.7b}$$

Dirichlet BCs are taken as

$$u(x = x_l, t) = g_2(t); \ \ u(x = x_u, t) = g_3(t) \tag{8.7c,d}$$

where $g_1(x), g_2(t), g_3(t)$ are prescribed functions.

An analytical solution is available for eqs. (8.7) for $\alpha = 2$ [1].

$$u_a(x, t) = \frac{1}{1 + \exp\left[\dfrac{x}{\sqrt{2D}} + (a - \dfrac{1}{2})t\right]} \tag{8.7e}$$

Equation (8.7e) can be used to define the IC and BCs. Note that $u_a(x,t)$ of eq. (8.7e) is a traveling wave solution with Lagrangian variable $\dfrac{x}{\sqrt{2D}} + (a - \dfrac{1}{2})t$. Since $a = 1$ for the following solutions $(a - \dfrac{1}{2}) > 0$ and the velocity of the traveling wave is negative (see the discussion of eqs. (8.6a,i) for further details).

$$u(x, t = 0) = g_1(x) = \frac{1}{1 + \exp\left[\dfrac{x}{\sqrt{2D}}\right]} \tag{8.7f}$$

$$u(x = x_l, t) = g_2(t) = \frac{1}{1 + \exp\left[\dfrac{x_l}{\sqrt{2D}} + (a - \dfrac{1}{2})t\right]} \tag{8.7g}$$

$$u(x = x_u, t) = g_3(t) = \frac{1}{1 + \exp\left[\dfrac{x_u}{\sqrt{2D}} + (a - \dfrac{1}{2})t\right]} \tag{8.7h}$$

8.6.1   MAIN PROGRAM

A main program for eqs. (8.7) follows.

Listing 8.9: Main program for eqs. (8.7).

```
#
# Fractional Fitzhugh-Nagumo
#
#    ut=D*(d^alpha u/dx^alpha)+u*(u-1)*(a-u)
#
#    xl < x < xu, 0 < t < tf, xl=-20, xu=10
#
#    u(x,t=0)=ua(x,t=0)
#
#    u(x=xl,t)=ua(x=xl,t)
#
#    u(x=xu,t)=ua(x=xu,t)
#
#    ua(x,t)=1/(1+exp(x/(2*D)^1/2+(a-1/2)*t))
#
# Delete previous workspaces
  rm(list=ls(all=TRUE))
#
# Access functions for numerical solution
```

```
library("deSolve");
setwd("f:/fractional/sfpde/chap8/ex5");
source("pde1a.R");
#
# Parameters
  ncase=1;
  if(ncase==1){a=1;D=1;   alpha=2;}
# if(ncase==1){a=1;D=0.1;alpha=2;}
# if(ncase==2){a=1;D=1;   alpha=1.5;}
# if(ncase==2){a=1;D=0.1;alpha=1.5;}
  sr=1/(sqrt(2*D));
#
# Boundary condition functions
  g_0=function(t) 1/(1+exp(xl/sqrt(2*D)+(a-0.5)*t));
  g_L=function(t) 1/(1+exp(xu/sqrt(2*D)+(a-0.5)*t));
#
#. Boundary condition coefficients
  c_1=function(t) 1;
  c_2=function(t) 0;
  c_3=function(t) 1;
  c_4=function(t) 0;
#
# Spatial grid
  xl=-60;xu=20;nx=101;dx=(xu-xl)/(nx-1);
  xj=seq(from=xl,to=xu,by=dx)+dx;
  cd=dx^(-alpha)/gamma(4-alpha);
  r12dx=1/(12*dx);
#
# Independent variable for ODE integration
  t0=0;tf=60;nt=4;dt=(tf-t0)/(nt-1);
  tout=seq(from=t0,to=tf,by=dt);
#
# a_jk coefficients
  A=matrix(0,nrow=nx-2,ncol=nx-1);
  for(j in 1:(nx-2)){
    for(k in 0:j){
    if (k==0){
      A[j,k+1]=(j-1)^(3-alpha)-j^(2-alpha)*(j-3+alpha);
    } else if (1 <= k && k<=j-1){
```

```
      A[j,k+1]=(j-k+1)^(3-alpha)-2*(j-k)^(3-alpha)+(j-k-1)^(3-
        alpha);
    } else
      A[j,k+1]=1;
    }
  }
#
# Initial condition
  u0=rep(0,nx-2);
  for(i in 1:(nx-2)){
    u0[i]=1/(1+exp(sr*xj[i+1]));
  }
  ncall=0;
#
# ODE integration
  out=lsodes(func=pde1a,y=u0,times=tout,
             sparsetype="sparseint")
  nrow(out)
  ncol(out)
#
# Allocate array for u(x,t)
  u=matrix(0,nt,nx);
#
# u(x,t), x ne xl,xu
  for(i in 1:nt){
    for(j in 2:(nx-1)){
      u[i,j]=out[i,j];
    }
  }
#
# Reset boundary values
  for(i in 1:nt){
      u[i,1]=1;
    u[i,nx]=0;
  }
#
# Display the order of the fractional derivatives
  cat(sprintf("\n    alpha = %4.2f\n",alpha));
#
```

```
# ncase=1
  if(ncase==1){
#
# Numerical, analytical solutions, difference
  uap=matrix(0,nt,nx);
  for(i in 1:nt){
    for(j in 1:nx){
      uap[i,j]=1/(1+exp(sr*xj[j]+(a-0.5)*tout[i]));
    }
  max_err=max(abs(u-uap));
  }
#
# Tabular numerical, analytical solutions,
# difference
  cat(sprintf("\n        t        x      u(x,t)    ua(x,t)        diff
      "));
  for(i in 1:nt){
  iv=seq(from=1,to=nx,by=10);
  for(j in iv){
    cat(sprintf("\n %6.2f%8.2f%10.5f%10.5f%12.3e",
      tout[i],xj[j],u[i,j],uap[i,j],u[i,j]-uap[i,j]));
  }
  cat(sprintf("\n"));
  }
#
# Plot numerical, analytical solutions
  matplot(xj,t(u),type="l",lwd=2,col="black",lty=1,
    xlab="x",ylab="u(x,t)",main="");
  matpoints(xj,t(uap),pch="o",col="black");
  }
#
# ncase=2
  if(ncase==2){
#
# Tabular numerical solution
  cat(sprintf("\n        t        x      u(x,t)"));
  for(i in 1:nt){
  iv=seq(from=1,to=nx,by=4);
  for(j in iv){
```

```
    cat(sprintf("\n %6.2f%8.2f%10.5f",
      tout[i],xj[j],u[i,j]));
  }
  cat(sprintf("\n"));
  }
#
# Plot numerical solution
  matplot(xj,t(u),type="l",lwd=2,col="black",lty=1,
    xlab="x",ylab="u(x,t)",main="");
  }
#
# Calls to ODE routine
  cat(sprintf("\n\n  ncall = %3d\n",ncall));
```

Listing 8.9 is similar to Listing 8.8. Therefore, only the differences are considered.

• Brief documentation is followed by the deletion of previous workspaces.

```
  #
  # Fractional Fitzhugh-Nagumo
  #
  #   ut=D*(d^alpha u/dx^alpha)+u*(u-1)*(a-u)
  #
  #   xl < x < xu, 0 < t < tf, xl=-20, xu=10
  #
  #   u(x,t=0)=ua(x,t=0)
  #
  #   u(x=xl,t)=ua(x=xl,t)
  #
  #   u(x=xu,t)=ua(x=xu,t)
  #
  #   ua(x,t)=1/(1+exp(x/(2*D)^1/2+(a-1/2)*t))
  #
  # Delete previous workspaces
    rm(list=ls(all=TRUE))
```

• Parameters are defined.

```
  #
  # Parameters
    ncase=1;
```

```
   if(ncase==1){a=1;D=1;   alpha=2;}
 # if(ncase==1){a=1;D=0.1;alpha=2;}
 # if(ncase==2){a=1;D=1;   alpha=1.5;}
 # if(ncase==2){a=1;D=0.1;alpha=1.5;}
   sr=1/(sqrt(2*D));
```

Basically, D, α in eq. (8.7a) are changed and a remains constant.

- The RHS functions of the BCs are functions of t defined by eqs. (8.7g,h).

```
 #
 # Boundary condition functions
   g_0=function(t) 1/(1+exp(xl/sqrt(2*D)+(a-0.5)*t));
   g_L=function(t) 1/(1+exp(xu/sqrt(2*D)+(a-0.5)*t));
```

- Dirichlet BCs are specified.

```
 #
 # Boundary condition coefficients
   c_1=function(t) 1;
   c_2=function(t) 0;
   c_3=function(t) 1;
   c_4=function(t) 0;
```

- The spatial grid of 101 points is defined over the interval $-60 \leq x \leq 20$ so the grid interval is $(20 - (-60))/100 = 0.8$ and $x = -60, -59.2, ..., 20$.

```
 #
 # Spatial grid
   xl=-60;xu=20;nx=101;dx=(xu-xl)/(nx-1);
   xj=seq(from=xl,to=xu,by=dx)+dx;
   cd=dx^(-alpha)/gamma(4-alpha);
   r12dx=1/(12*dx);
```

- The interval in t is $0 \leq t \leq 60$ with output values $t = 0, 20, 40, 60$.

```
 #
 # Independent variable for ODE integration
   t0=0;tf=60;nt=4;dt=(tf-t0)/(nt-1);
   tout=seq(from=t0,to=tf,by=dt);
```

- IC (8.7f) is programmed as

```
#
# Initial condition
  u0=rep(0,nx-2);
  for(i in 1:(nx-2)){
    u0[i]=1/(1+exp(sr*xj[i+1]));
  }
  ncall=0;
```

The counter for the calls to ODE/MOL routine pde1a is also initialzed.

- The integration of the $101 - 2 = 99$ ODEs at the interior points in x is by lsodes.

```
#
# ODE integration
  out=lsodes(func=pde1a,y=u0,times=tout,
             sparsetype="sparseint")
  nrow(out)
  ncol(out)
```

out has dimensions $4 \times 99 + 1 = 100$ as indicated in the solutions that follow.

- After placing the numerical solution in array u, the boundary values are set (by *a priori* knowledge of the solutions).

```
#
# Reset boundary values
  for(i in 1:nt){
      u[i,1]=1;
      u[i,nx]=0;
  }
```

- For ncase=1, the numerical and analytical solutions and the difference are displayed.

```
#
# ncase=1
  if(ncase==1){
#
# Numerical, analytical solutions, difference
  uap=matrix(0,nt,nx);
  for(i in 1:nt){
    for(j in 1:nx){
```

```
          uap[i,j]=1/(1+exp(sr*xj[j]+(a-0.5)*tout[i]));
        }
    max_err=max(abs(u-uap));
    }
#
# Tabular numerical, analytical solutions,
# difference
    cat(sprintf("\n      t       x     u(x,t)   ua(x,t)     diff"));
    for(i in 1:nt){
    iv=seq(from=1,to=nx,by=10);
    for(j in iv){
      cat(sprintf("\n %6.2f%8.2f%10.5f%10.5f%12.3e",
        tout[i],xj[j],u[i,j],uap[i,j],u[i,j]-uap[i,j]));
    }
    cat(sprintf("\n"));
    }
```

• For ncase=1, the numerical solution is plotted with `matplot` and the analytical solution is superimposed with `matpoints`.

```
#
# Plot numerical, analytical solutions
    matplot(xj,t(u),type="l",lwd=2,col="black",lty=1,
      xlab="x",ylab="u(x,t)",main="");
    matpoints(xj,t(uap),pch="o",col="black");
    }
```

The transposes `t(u)`, `t(uap)` are required so that the dimensions of `xj`, `u`, `ua` conform (are consistent). The solutions are plotted parametrically in t.

The final } concludes ncase=1.

• For ncase=2, only the numerical solution is displayed.

```
#
# ncase=2
    if(ncase==2){
#
# Tabular numerical solution
    cat(sprintf("\n      t       x     u(x,t)"));
    for(i in 1:nt){
```

```
iv=seq(from=1,to=nx,by=4);
for(j in iv){
  cat(sprintf("\n %6.2f%8.2f%10.5f",
    tout[i],xj[j],u[i,j]));
}
cat(sprintf("\n"));
}
```

- The numerical solution is plotted.

```
#
# Plot numerical solution
  matplot(xj,t(u),type="l",lwd=2,col="black",lty=1,
    xlab="x",ylab="u(x,t)",main="");
  }
```

- At the end of the solutions, the number of calls to pde1a is displayed.

```
#
# Calls to ODE routine
  cat(sprintf("\n\n  ncall = %3d\n",ncall));
```

This completes the discussion of the programming of the main program in Listing 8.9.

8.6.2 ODE/MOL ROUTINE

The ODE/MOL routine pde1a called by lsodes is similar to the routines in Listings 8.2 and 8.6 and therefore is not listed here. The programming of eq. (8.7a) is the essential difference.

```
ut[j]=cd*D*ut[j]+u[j]*(u[j]-1)*(a-u[j]);
```

The programming of the nonlinear source term $u(u - 1)(a - u)$ is noteworthy.

The numerical and graphical output for ncase=1,2 in the main program of Listing 8.9 is reviewed next.

8.6.3 MODEL OUTPUT

We can note the following details about the output in Table 8.16.

- The dimensions of the solution matrix out from lsodes are $4 \times 99 + 1 = 100$ for the 99 ODEs at the interior points in x plus the offset of +1 to include the value of t as the first element of each of the 4 solution vectors.

Table 8.16: Numerical solution for eqs. (8.7), ncase=1, $a = 1, D = 1, \alpha = 2$

[1] 4

[1] 100

alpha = 2.00

t	x	u(x,t)	ua(x,t)	diff
0.00	-60.00	1.00000	1.00000	0.000e+00
0.00	-52.00	1.00000	1.00000	0.000e+00
0.00	-44.00	1.00000	1.00000	0.000e+00
0.00	-36.00	1.00000	1.00000	0.000e+00
0.00	-28.00	1.00000	1.00000	0.000e+00
0.00	-20.00	1.00000	1.00000	0.000e+00
0.00	-12.00	0.99979	0.99979	0.000e+00
0.00	-4.00	0.94419	0.94419	0.000e+00
0.00	4.00	0.05581	0.05581	0.000e+00
0.00	12.00	0.00021	0.00021	0.000e+00
0.00	20.00	0.00000	0.00000	-7.214e-07

.
.
.   .
.

Output for t = 20, 40, removed

.   .
.   .
.   .

60.00	-60.00	1.00000	1.00000	4.011e-06
60.00	-52.00	0.99881	0.99885	-4.376e-05
60.00	-44.00	0.76926	0.75263	1.663e-02
60.00	-36.00	0.01147	0.01052	9.482e-04
60.00	-28.00	0.00004	0.00004	5.417e-06
60.00	-20.00	0.00000	0.00000	2.482e-08
60.00	-12.00	0.00000	0.00000	1.114e-10
60.00	-4.00	0.00000	0.00000	4.773e-13
60.00	4.00	0.00000	0.00000	2.018e-15
60.00	12.00	0.00000	0.00000	8.049e-18
60.00	20.00	0.00000	0.00000	-6.750e-20

ncall = 351

```
[1] 4

[1] 100
```

- For ncase=1, $\alpha = 2$

  ```
  alpha = 2.00
  ```

- The IC for the numerical and analytical solutions is the same since they are set by the analytical solution of eqs. (8.7b,f).

- At $t = 0$, a sharp front occurs within the interval between $x = -4$ and $x = 4$.

  ```
  0.00    -4.00    0.94419    0.94419    0.000e+00
  0.00     4.00    0.05581    0.05581    0.000e+00
  ```

- At $t = 60$, this front moved to the interval between $x = -44$ and $x = -36$.

  ```
  60.00   -44.00    0.76926    0.75263    1.663e-02
  60.00   -36.00    0.01147    0.01052    9.482e-04
  ```

  That is, the movement is right to left. This movement occurs even with just parabolic diffusion in eq. (8.7a) (with $\alpha = 2$) as a result of the nonlinear source term.

- The largest error is for $u(x = -44, t = 60)$ but this does not invalidate the solution since the error occurs where the solution is changing rapidly (at the front). This point is demonstrated in Fig. 8.17 (the additional points in Fig. 8.17 appear because the solutions are plotted for all 101 values of x while in the numerical output above, the solutions are displayed for every tenth value of x, so the spacing between the values of x is $(10)(20 - (-60))/100 = 8)$.

  ```
  1.663e-02
  ```

- The computational effort is modest ncall = 351.

  For ncase=1, $a = 1, D = 0.1, \alpha = 2$ (D is reduced from 1 to 0.1), the dispersion (diffusion) in x is reduced. This reduced dispersion is reflected in the numerical solution in Table 8.17. The front at $t = 60$ is now confined to the interval $x = -20$ to $x = -4$.

  ```
  60.00   -20.00    1.00000    1.00000   -1.729e-06
  60.00   -12.00    0.07750    0.04042    3.708e-02
  60.00    -4.00    0.00000    0.00000    3.927e-09
  ```

Table 8.17: Numerical solution for eqs. (8.7), ncase=1, $a = 1, D = 0.1, \alpha = 2$

[1] 4

[1] 100

alpha = 2.00

t	x	u(x,t)	ua(x,t)	diff
0.00	-60.00	1.00000	1.00000	0.000e+00
0.00	-52.00	1.00000	1.00000	0.000e+00
0.00	-44.00	1.00000	1.00000	0.000e+00
0.00	-36.00	1.00000	1.00000	0.000e+00
0.00	-28.00	1.00000	1.00000	0.000e+00
0.00	-20.00	1.00000	1.00000	0.000e+00
0.00	-12.00	1.00000	1.00000	0.000e+00
0.00	-4.00	0.99987	0.99987	0.000e+00
0.00	4.00	0.00013	0.00013	0.000e+00
0.00	12.00	0.00000	0.00000	0.000e+00
0.00	20.00	0.00000	0.00000	-3.782e-20

```
            .              .
            .              .
            .              .

    Output for t = 20, 40, removed

            .              .
            .              .
            .              .
```

60.00	-60.00	1.00000	1.00000	0.000e+00
60.00	-52.00	1.00000	1.00000	0.000e+00
60.00	-44.00	1.00000	1.00000	0.000e+00
60.00	-36.00	1.00000	1.00000	0.000e+00
60.00	-28.00	1.00000	1.00000	-7.916e-13
60.00	-20.00	1.00000	1.00000	-1.729e-06
60.00	-12.00	0.07750	0.04042	3.708e-02
60.00	-4.00	0.00000	0.00000	3.927e-09
60.00	4.00	0.00000	0.00000	2.551e-16
60.00	12.00	0.00000	0.00000	1.517e-23
60.00	20.00	0.00000	0.00000	-3.539e-33

ncall = 373

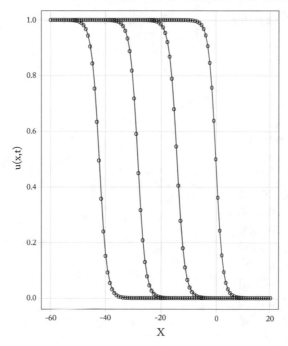

Figure 8.17: Numerical solution of eqs. (8.7), ncase=1, $a = 1, D = 1, \alpha = 2$.

These features of the solution are also reflected in Fig. 8.18.

Even with the front sharpening (from $D = 1$ to $D = 0.1$), the agreement between the numerical and analytical solutions is acceptable.

For ncase=2, $a = 1, D = 1, \alpha = 1.5$, the order of the fractional derivative in eq. (8.7a) is reduced to closer to 1 (from parabolic for $\alpha = 2$ to hyperbolic-parabolic for $\alpha = 1.5$). This reduction in α adds additional movement of the front from right to left, and in fact, it reaches the left boundary. Therefore, the left boundary is extended from $x_l = -60$ to $x_l = -80$.

```
#
# Spatial grid
  xl=-80;xu=20;nx=101;dx=(xu-xl)/(nx-1);
  xj=seq(from=xl,to=xu,by=dx);
```

Abbreviated numerical output is shown in Table 8.18 and the graphical output in Fig. 8.19.

The front at $t = 60$ is now approximately between $x = -60$ and $x = -50$.

```
  60.00  -60.00    0.71555
  60.00  -50.00    0.01605
```

These features of the solution are also reflected in Fig. 8.19.

Table 8.18: Numerical solution for eqs. (8.7), ncase=2, $a = 1, D = 1, \alpha = 1.5$

```
[1]  4

[1]  100

   alpha = 1.50

       t        x      u(x,t)
    0.00   -80.00    1.00000
    0.00   -70.00    1.00000
    0.00   -60.00    1.00000
    0.00   -50.00    1.00000
    0.00   -40.00    1.00000
    0.00   -30.00    1.00000
    0.00   -20.00    1.00000
    0.00   -10.00    0.99915
    0.00     0.00    0.50000
    0.00    10.00    0.00085
    0.00    20.00    0.00000
                .
                .
                .
      Output for t = 20,
         40, removed
                .
                .
                .
   60.00   -80.00    1.00000
   60.00   -70.00    0.99987
   60.00   -60.00    0.71555
   60.00   -50.00    0.01605
   60.00   -40.00    0.00402
   60.00   -30.00    0.00199
   60.00   -20.00    0.00124
   60.00   -10.00    0.00087
   60.00     0.00    0.00065
   60.00    10.00    0.00051
   60.00    20.00    0.00000

   ncall = 1020
```

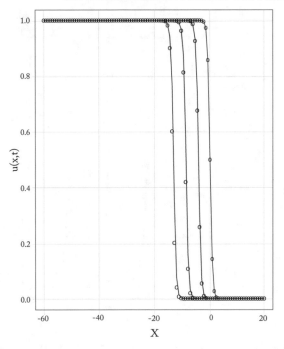

Figure 8.18: Numerical solution of eqs. (8.7), `ncase=1`, $a = 1, D = 0.1, \alpha = 2$.

As discussed previously (for the main program of Listing 8.8), an analytical solution is not included.

For `ncase=2`, $a = 1, D = 0.1, \alpha = 1.5$, the dispersion (diffusion) of Fig. 8.19 is reduced, as reflected in the abbreviated numerical output in Table 8.19 and in Fig. 8.20.

The front at $t = 60$ is now approximately between $x = -20$ and $x = -10$.

```
60.00   -20.00    1.00000
60.00   -10.00    0.00717
```

These features of the solution are also reflected in Fig. 8.20.

The spatial resolution with 101 points appears to be adequate for this case of a sharp front (there are no obvious numerical distortions such as diffusion and oscillation). Therefore, the preceding combination of the spline approximation of the fractional derivative of eq. (8.7a) and the MOL appears to give acceptable numerical solutions. This could be further investigated by varying `nx=101` (h refinement).

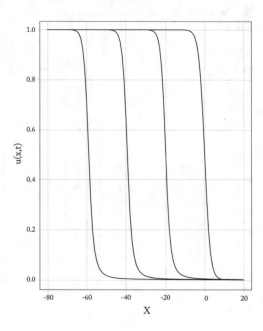

Figure 8.19: Numerical solution of eqs. (8.7), `ncase=2`, $a = 1, D = 1, \alpha = 1.5$.

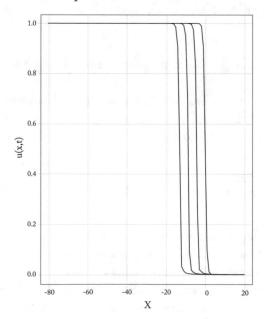

Figure 8.20: Numerical solution of eqs. (8.7), `ncase=2`, $a = 1, D = 0.1, \alpha = 1.5$.

Table 8.19: Numerical solution for eqs. (8.7), ncase=2, $a = 1, D = 0.1, \alpha = 1.5$

[1] 4

[1] 100

```
  alpha = 1.50

     t       x     u(x,t)
  0.00  -80.00   1.00000
  0.00  -70.00   1.00000
  0.00  -60.00   1.00000
  0.00  -50.00   1.00000
  0.00  -40.00   1.00000
  0.00  -30.00   1.00000
  0.00  -20.00   1.00000
  0.00  -10.00   1.00000
  0.00    0.00   0.50000
  0.00   10.00   0.00000
  0.00   20.00   0.00000

              .
              .
              .

    Output for t = 20,
       40, removed

 60.00  -80.00   1.00000
 60.00  -70.00   1.00000
 60.00  -60.00   1.00000
 60.00  -50.00   1.00000
 60.00  -40.00   1.00000
 60.00  -30.00   1.00000
 60.00  -20.00   1.00000
 60.00  -10.00   0.00717
 60.00    0.00   0.00062
 60.00   10.00   0.00026
 60.00   20.00   0.00000

 ncall = 958
```

8.7   SUMMARY AND CONCLUSIONS

This chapter demonstrates through a series of example applications for the extension of integer PDEs to SFPDEs. The common basis for these examples is: (1) the spline (or finite element) approximation of the fractional derivatives and (2) the MOL approximation of the SFPDEs by a system of ODEs that can then be integrated numerically by a library initial value, ODE integrator, e.g., lsode, lsodes. The Fisher-Kolmogorov equation discussed in Chapter 5 is another example of an integer to SFPDE extension.

Other integer PDEs that could be considered for extension to SFPDEs include integer PDEs: (1) second order in t, (2) in 2D and 3D, and (3) in coordinate systems other than Cartesian, e.g., cylindrical, spherical.

The details for extensions of these integer PDEs to SFPDEs could, in principle, follow the procedures discussed previously based on the use of approximations for the fractional derivatives in combination with the MOL. These additional types of integer PDEs can serve as the basis for future computer implementation and application studies of SFPDEs. The numerical methods and R routines discussed in this book can possibly serve as a starting point for these additional studies.

REFERENCES

[1] Griffiths, G.W. and W.E. Schiesser (2012), *Traveling Wave Analysis of Partial Differential Equations*, Elsevier, Burlington, MA, Chapter 9. 350

[2] Madsen, N.K. and R.F. Sincovec (1976), General Software for Partial Differential Equations, in *Numerical Methods for Differential Systems*, L. Lapidus and W.E. Schiesser (eds.), New York, San Diego, 1976. 294

[3] Schiesser, W.E. (1991), *The Numerical Method of Lines Integration of Partial Differential Equations*, Academic Press, San Diego. 308

Authors' Biographies

YOUNES SALEHI

My research focus is applied mathematics broadly. This includes numerical linear algebra, optimization and solving differential equations. My primary research interest concerns the areas of numerical analysis, scientific computing and high performance computing with particular emphasis on the numerical solution of ordinary differential equations (ODEs) and partial differential equations (PDEs).

One focus of my work is programming efficient numerical methods for ODEs and PDEs. I have extensive experience in MATLAB, Maple, Mathematica and R programming of transportable numerical method routines, but I am also experienced in programming in C, C++ and C#, and could readily apply these programming systems to numerical ODE/PDEs.

Recently, I have become interested in fractional differential equations (FDEs), especially the numerical solution of fractional initial value problems (FIVPs) and space fractional differential equations (SFPDEs).

WILLIAM E. SCHIESSER

William E. Schiesser is Emeritus McCann Professor of Computational Biomedical Engineering and Chemical and Biomolecular Engineering, and Professor of Mathematics at Lehigh University. His research is directed toward numerical methods and associated software for ordinary, differential-algebraic and partial differential equations (ODE/DAE/PDEs). He is the author, coauthor or coeditor of 18 books, and his ODE/DAE/PDE computer routines have been accessed by some 5,000 colleges and universities, corporations and government agencies.

Index

Printed in the United States
by Baker & Taylor Publisher Services